The Game Music Toolbox

AF766733

The Game Music Toolbox provides readers with the tools, models, and techniques to create and expand a compositional toolbox, through a collection of 20 iconic case studies taken from different eras of game music. Discover many of the composition and production techniques behind popular music themes from games such as *Cyberpunk 2077*, *Mario Kart 8*, *The Legend of Zelda*, *Street Fighter II*, *Diablo*, *Shadow of the Tomb Raider*, *The Last of Us*, and many others.

The Game Music Toolbox features:

- Exclusive interviews from industry experts
- Transcriptions and harmonic analyses
- 101 music theory introductions for beginners
- Career development ideas and strategies
- Copyright and business fundamentals
- An introduction to audio implementation for composers
- Practical takeaway tasks to equip readers with techniques for their own game music

The Game Music Toolbox is crucial reading for game music composers and audio professionals of all backgrounds, as well as undergraduates looking to forge a career in the video game industry.

Marios Aristopoulos is a composer and sound designer for new media based in London. He is the Game Audio Lead at Guildhall School of Music and Drama and has authored the game audio programmes for many universities in the UK and the USA. Selected composition credits include *Apotheon* – an award-winning PS4 & Steam video game, *Rebel Rebel* – a painting exhibition for The Curve gallery in the Barbican, *Aenigma* – a 3D stereoscopic animation, and *Beasts of London* – an interactive exhibition for the Museum of London.

The Game Music Toolbox

Composition Techniques and Production Tools from 20 Iconic Game Soundtracks

Marios Aristopoulos

Routledge
Taylor & Francis Group
LONDON AND NEW YORK

Designed cover image: Sashatigar/Shutterstock.com

First published 2023
by Routledge
4 Park Square, Milton Park, Abingdon, Oxon OX14 4RN

and by Routledge
605 Third Avenue, New York, NY 10158

Routledge is an imprint of the Taylor & Francis Group, an informa business

© 2023 Marios Aristopoulos

The right of Marios Aristopoulos to be identified as author of this work has been asserted in accordance with sections 77 and 78 of the Copyright, Designs and Patents Act 1988.

All rights reserved. No part of this book may be reprinted or reproduced or utilised in any form or by any electronic, mechanical, or other means, now known or hereafter invented, including photocopying and recording, or in any information storage or retrieval system, without permission in writing from the publishers.

Trademark notice: Product or corporate names may be trademarks or registered trademarks, and are used only for identification and explanation without intent to infringe.

British Library Cataloguing-in-Publication Data
A catalogue record for this book is available from the British Library

Library of Congress Cataloging-in-Publication Data
Names: Aristopoulos, Marios, author.
Title: The game music toolbox : composition techniques and production tools from 20 iconic game soundtracks / Marios Aristopoulos.
Description: New York: Routledge, 2023. | Includes bibliographical references and index.
Subjects: LCSH: Video game music—Instruction and study. | Video game music—Analysis, appreciation.
Classification: LCC MT64.V53 A75 2023 (print) | LCC MT64.V53 (ebook) | DDC 781.5/4—dc23/eng/20221115
LC record available at https://lccn.loc.gov/2022055446
LC ebook record available at https://lccn.loc.gov/2022055447

ISBN: 978-0-367-70550-3 (hbk)
ISBN: 978-0-367-70549-7 (pbk)
ISBN: 978-1-003-14687-2 (ebk)

DOI: 10.4324/9781003146872

Typeset in Sabon
by codeMantra

Access the Support Material: www.routledge.com/9780367705497

Contents

Preface

Welcome to the Game Music Toolbox!

Aims of the book

The purpose of this book is to expand your compositional toolbox with inspirational techniques taken from 20 iconic game soundtrack of different historical periods, introduce you to some of the production gear and technology used by AAA and indie game composers, and offer you practical information on how to forge your own career path in the industry. The ideas discussed are not meant to provide a strict compositional framework but rather function as starting points for your own artistic journey. You do not have to necessarily use any of the techniques in the exact same way as the original composers, but you can transform them into something new.

How to use this book

This book is structured like an open world video game, you can navigate it in any order you want! The introduction is designed to provide you with practical and theoretical information on different aspects of a career in game composition in a Q&A friendly format, like a tutorial you can go through before jumping into the main adventure. It is only natural that you might already be more familiar with some of the material, feel free to skip anything you already know; I guarantee that you will find something new.

Each of the book's 20 main chapters is based on a soundtrack case study arranged chronologically, starting at the age of the arcades with *Space Invaders* in 1978 and reaching today's age of cloud gaming with *Cyberpunk 2077*. Chapters begin with a brief background information on the game and the composer, with an emphasis on how the composer got involved in the project. The main part of each chapter presents one or more composition techniques used in the chosen game and a selection of the production tools utilized by the composer. As everyone has different strengths and weaknesses in their compositional toolkit, I have occasionally included

101-theory sections to help bridge any potential knowledge gaps you might have. At the end of each chapter, you will find hands-on takeaway tasks you can complete, ranked by difficulty. These are designed to help you apply these new ideas into your own music, and you can slightly modify the details if the core idea of the exercise is maintained. There is also a bibliography for each chapter to indicate the sources of the ideas discussed.

Finally, to benefit the most from reading this book, make sure to follow the accompanying video examples as they are an integral part of understanding each case study. You might have to navigate through different parts of the video as indicated by the timecode in the text, but I have left the videos unedited so you can examine other parts as you like.

Link to Video Examples Playlist can be found here:
www.routledge.com/9780367705497

About me and the team behind the book

Since 2016, I have been the Game Audio Lead professor at Guildhall School of Music & Drama in London which ranked as the top university in the UK for studying Music in 2023 by the Guardian University Guide music league table. I have previously held senior music lecturer positions at BIMM London, Point Blank London, The Institute of Audio Research in New York City, and the Greek National Conservatory in Athens. I hold a PhD in game audio from City University, a Masters in Acoustic composition, a second Masters in Ethnomusicology from SOAS, and a Bachelor's in music from Goldsmiths. I have presented my game audio research in multiple international conferences such as Ludomusicology, MaMI in NYU Steinhardt, and the American Musicological Society/SMT.

Aside from my academic career, I have many years of industry experience as a composer and sound designer for new media. My music for the video game *Apotheon* was selected as one of the top 10 PlayStation 4 soundtracks of 2016, and I was extremely honoured that it was featured in the Olympic Flame ceremony of the 2019 Special Olympics in Greece. I have worked with the London-based studio Gram Games (owned by Zynga), the Chinese studio Youzoo games, and the French studio Deepnight games. Outside games, I frequently work in animation, theatre, and film in the USA and Europe. You can see examples of my work at mariosaristopoulos.com.

While writing this book, I had the benefit of collaborating with two amazing researchers and composers, Edie Evans and James Allen, who contributed to the musicological analysis and transcriptions of the case studies presented in this book. Most of the information presented in each case study is based on the original composer interviews from numerous sources that are indicated at the end of each chapter. However, I also had the pleasure of having fascinating conversations directly with some of the original composers who kindly agreed to be interviewed exclusively for this book (Wilbert Roget II, Petri Alanko, Paul Weir, and Lorenzo Bassignani).

Introduction

15 Questions on game composition

This extended introduction is based on the most frequently asked questions I have received by my students on different aspects of game composition. The questions are organized into five categories: scoring process, implementation, career, business and money, and education. My answers are based on my own industry experience and observations, conversations with other members of the industry, and data from extended composer surveys quoted at the end of the chapter.

Scoring process

What is the purpose of game music?

One way of clarifying the contribution of a musical piece in a video game is to simply play it with the original music turned off and observe how the experience changes. A fantasy game like *Zelda* might feel less adventurous without its magical soundtrack; a survival horror game like *Resident Evil* might feel less terrifying without its clever musical jump scares; while a music game such as *Guitar Hero* will simply be unplayable as music is a fundamental game mechanic. While the purpose of music can greatly vary from one game to another, we can observe that it usually serves one or more of the following functions.

It adds entertainment value

First, and most importantly, good game music makes games more fun to play. Isn't fun what it is supposed to be what games are all about? There might be some types of educational or competitive games in which player entertainment is not the primary goal but for most games adding the right type of music can simply make playing more enjoyable and satisfying.

DOI: 10.4324/9781003146872-1

It provides information

Music can be used as an abstract tool for directly communicating information to the player in many creative ways (see Chapter 13: *Mario Kart*). An example is the use of short music stingers that notify the player about a particular change in a game, while a sudden change in the music is usually suggestive of a switch of gameplay state that might have gone unnoticed (eg., the use of battle music to warn about the presence of incoming enemies). Music can also be used to evoke a particular time and place through association with different musical cultures (see Chapter 9: *Assassin's Creed*).

It evokes different moods

The power of music to produce and enhance a wide range of moods is universally recognized and is one of the most central functions of music in media. The effectiveness of evoking a particular mood is influenced by many factors including personal taste, but its communication can be relatively easy to identify as researchers have proven that most listeners will effectively distinguish between musical pieces that aim to portray different basic emotions (ex: ear, joy, sadness, surprise).[1] Game composers frequently use music as an abstract emotional language to enhance storytelling and provide hints of the emotional state of characters (see Chapter 11: *The Last of Us*).

It assists with memory and understanding of the game world

Associating a piece of music with a particular gameplay context can be helpful in reducing the learning curve for new players. For example, in the multiplayer game *World of Warcraft*, the use of location music helped new players recognize and distinguish important areas as they navigated through an open world consisting of hundreds of locations.[2] I can also anecdotally state that good game music creates a memorable identity that can extend the gaming experience outside playing time. It is hard to think of my favourite games without their soundtrack coming to mind.

It enhances immersion

Gameplay immersion is something difficult to measure accurately but it can be argued that successful use of music can have a positive impact. On a superficial level, music can amplify immersion by simply covering any sounds originating outside the game (outside traffic, loud conversations, the sound of your graphics card overheating!) and help you focus on what is happening in the game. On a deeper level, a successful soundtrack might assist in creating a more meaningful and enjoyable experience that will naturally generate higher levels of engagement.

How do you choose the right musical style for a game?

The musical language of video games has been incredibly diverse over the past 40 years: from chiptune (*Casltevania*) to early jazz (*Cuphead*), metalcore (*Doom*), orchestral (*Final Fantasy*), 12 tone (*Metamorphosis*), industrial noise (*Quake*), EDM (*Mirror's Edge*), hip hop (*Need for Speed*), found sound (*Control*), rock (*guitar hero*), Japanese traditional music (*Ghost of Tsushima*); the list goes on and on as game music encompasses almost every musical style imaginable. Choosing an appropriate aesthetic for a game can therefore be a challenge but is one of the most important decisions to be taken by the composer as it can have a strong impact on both the playing experience and the game's success. Unfortunately, there is no magic formula; this is a personal artistic question that each composer must answer in their own unique way. Hopefully, the range of case studies in this book will give you a starting point but here are some influential factors that I suggest taking into consideration while forging your musical vision for a game:

- What is the **thematic** and narrative context of the game? Pirates, ghosts, sci-fi, post-apocalyptic? Serving the story and setting of the game should be one of the primary factors to consider.
- What is the **visual style**? Cartoonish, pixel art, realistic, polygon? The graphics and art of the game can be a major source of inspiration. For example, a game using pixelated graphics and a limited colour palette could call for the music to also incorporate some level of audio limitations of the 8-bit retro era. These could range from a full-on tracker-based sequencer using basic square waves and PSGs (see Chapters 1 and 4) to using a Bitcrusher effect (see Chapter 8).
- What is the **gaming genre**? Platformer, FPS, strategy, adventure, sports simulator? Gaming genres tend to follow a particular design structure that can influence the musical structure. For example, a turn-based strategy game like *Total War* franchise has a map mode, a battle mode, and various events/cinematics. Observing the structure of the game will be helpful in determining what type of music is needed in each part.
- What is the overall **gameplay intensity**? Relaxed, casual, hectic? This is an area that composers can easily misinterpret, especially if they are not gamers themselves. The usual approach of a film composer might be to write intense music for an intense scene. This might work in a gaming context for the short term, but because of the overload of sensory information that occurs during active gameplay, the addition of intense music for prolonged time durations can easily overwhelm the player. This is an interesting relationship to examine but some action-heavy games (ex: *Starcraft II*, or *Diablo*) tend to take the opposite approach and introduce less or lighter music during combat or even no music at all (ex: *Fortnite*, *Overwatch*).

- What is the **platform's audio system**? Mobile phone, PC speakers, home TV, VR headset? This is a smaller factor, but the gaming platform can have an effect on how the music is experienced. For example, if a game is designed exclusively for mobile phones, then having an orchestral score with a wide dynamic range might not work very effectively, and a lot of the music might need to be compressed. Similarly, a VR game will most likely be experienced over headphones so this should be considered during the production of the music.
- What are the **audience expectations** for this type of game and what has been done before? You can always play it safe and build upon past trends, or you can attempt to break new ground and experiment with something original, the choice is yours! The risk of trying something completely different is that it might not register effectively to a general audience that is used to certain past trends. For example, using lightly distorted electric guitars for a Western type of game (ex: *Red Dead Redemption*) is a cliché which is completely anachronistic – there were no electric guitars in the far West! Such clichés work very well because we are conditioned through the great music of the past to recognize it as a distinguishing feature of that era. However, just repeating past traditions without trying anything new is a recipe for boring gaming music.

What is the process of scoring for games? Is it that different than film?

The process of scoring a video game can be quite different from other linear media such as films. Depending on the game company and the individual project, a composer might be involved from the very beginning of this process or might be brought in much later. You can scroll through video example 1 to observe how drastically the AAA game *Horizon Zero Dawn* changed from early prototyping stages through multiple years of production. Here are the typical stages of game development and the general uses of music in each.

Concept/pre-production stage

The point of this early stage is to establish an overall vision of how the game will feel, look, and sound like and to develop an early working prototype to test these ideas. It is quite common for developers to also use reference music from pre-existing games or other media as a temporary placeholders (video example 1, 11:08). Temp tracks can be a useful communication tool as not everyone is proficient with using appropriate music terminology to express their ideas. Moreover, they can also offer a way for composers to be discovered by game companies as temp music can often end up being licensed for the final product.

If a composer is brought on into a project at this early stage, she or he will likely begin writing music based mainly on concept art and game design documents, along conversations with the game design team, as not much of a playable game would exist. Learning how to work with concept art is a very useful skill for game composers (see task 3).

Production stage

Games will change significantly during the production stage as more assets are added or removed. At this stage of development, if composers have compatible hardware and software and feel confident enough to take on a gaming challenge, they can playtest working parts of the game to inform and inspire their music writing. There is no shame in asking programmers for shortcut mechanisms, such as cheats, to make skipping through parts of the game easier and faster. For the less daring, it is also very common to ask for gameplay video captures of different levels to help speed up the process. It takes some imagination to understand how a game might end up looking and feeling, and big changes can happen fast, so a great amount of flexibility, speed, and patience is required from the composer at this stage.

Just imagine the following hypothetical situation that is not that far-fetched from reality: You just completed writing a theme based on your play-through of a certain peaceful level. Then you play it again after three weeks, but now everything is on fire, it lasts ten times as long, there is a boss enemy, and puzzles, and it no longer happens at the beginning of the game but rather at the end! Frustrated you change the music almost completely. You then play the game again after another three months, but now the level does not exist anymore, they have taken it out! It is not unheard of for entire levels to be scrapped or important characters to be replaced that can impact the music direction in the very last minute and that is why being highly flexible is crucial. Composers might not always admit this publicly, but a piece might be originally written for an entirely different context than the one that it ended up being used for. As an example, the Oscar winning composer Trent Reznor wrote the soundtrack for the original 1996 *Quake* as an album and then the developers arranged the music on different levels as they saw fit.[3]

Beta testing

When a game is polished enough to be playable, it enters the beta testing stage in which it is publicly opened to a select number of players (depending on if it is an open or closed beta), that will play the game and provide feedback before the official release. Usually, changes at this stage are relatively small and are limited to polishing the experience and fixing bugs. Composers might start working on the trailer music as well as making a promotional soundtrack album to be released right before launch.

Post-release

As opposed to a film that is fixed and finalized upon release, many games continue to be updated after launch. In some cases, these updates do not affect the music, and the composer's job is done. However, big game franchises such as *Assassins Creed* or *Witcher* usually introduce new expansion packs that enrich the original content with additions to the story that usually require new music, while the old part of the game is largely left unchanged. These types of expansions are generally written by the same team of composers to maintain continuity with the original title. What can be even more challenging are online games (ex: MMORPGs) with monthly subscriptions that are in perpetual development. New content might be added (or removed) constantly to keep players engaged which might impact the use of music over time. Think of an online game like *Eve Online* which was released in 2003 and is still updated regularly with new content that now includes over 7,800 different star systems that can be visited by online players. As the universe expanded over time so did the music needs, with new cues and interactive mechanisms added, but also old music loved by fans had to be removed together with the old level designs.

Implementation

What is interactive music?

If the timing and order of all possible player actions that can occur in a game would be completely known, then we could produce a predetermined game soundtrack in the same way we do with films. However, the ability of players to partially influence the development of a game through some type of input (ex: through a controller) introduces a very large number of possible gameplay variations that can potentially exist. The differences between each possible outcome can vary from subtle differences in the timing of the same events that unfold linearly, to drastically different story paths in an open-ended game. If we want the music to remain as relevant as possible to every variation, we need to use an interactive music system that can be altered to some degree in response to adapt to gameplay changes. The well-known ludomusicologist (meaning game music academic!) Karen Collins differentiates between the terms of interactive, adaptive, and dynamic music, to describe the specific ability of the system to change due to direct player input (interactive), pre-determined game events (adaptive), or both (dynamic). In the gaming industry, most composers and designers use these three terms interchangeably with the most common one being interactive.

To understand this concept better let us take a very simple game like *Pac-Man* and hypothesize how music could be used interactively (Figure 0.1).

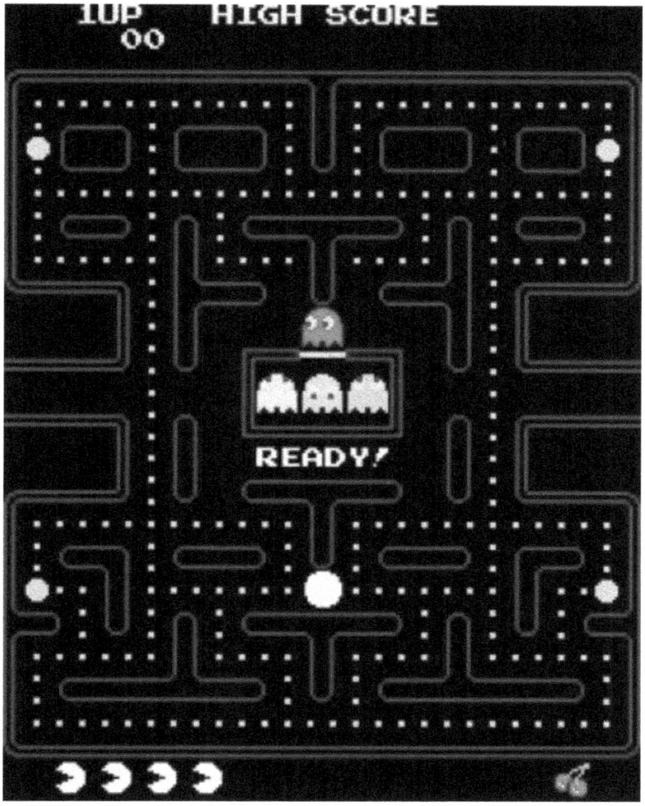

Figure 0.1 A gameplay screenshot from the original Pac-Man (1980). The goal of the game is to eat all the dots inside the maze while simultaneously avoiding the four colourful ghosts.

What are the different gameplay variations that this simple game can potentially generate? The player is given limited movement options within the maze (controlled by a joystick) but the possible pathways in combination with the timing variables of each change in direction would suggest that there is perhaps an infinite number of possible variations. However, does the music really need to be that different for each? An interactive music enthusiast might argue that we could build a sophisticated generative system produces a real-time soundtrack according to parameters from the game (ex: number of ghosts remaining). Another composer might prefer to define the main game events and then use a suitable music cue for each: a happy theme for when the level is completed, a sad theme for when the player loses the game, and a special theme for when the player temporarily becomes invincible and chases the ghosts around. Finally, another composer might argue that the best approach is to create a looped playlist of catchy and fun songs that are unaffected by gameplay to help relax the player. All these approaches might be suitable despite their differences in technical complexity, and this is a creative and artistic choice that is part of the process of interactive composition.

Does good game music need to be highly interactive?

Fortunately, producing an interactive music system that can work with a near infinite number of gameplay variations is not necessarily as complicated as it sounds for two reasons. First, music is an abstract art form so usually the same piece of music (or a slight variation of it) can be suitable with many gameplay variations without contradicting the experience. This is especially true if the differences of each playthrough are small as we saw in the *Pac-man* example. Second, the interactive boundaries of most games, including games that allow greater player autonomy, are still relatively limited and easily predictable. Players usually operate within a narrow amount of freedom as they are constrained to a pre-determined number of possible actions. Perhaps this might change in the future as interactive storytelling becomes increasingly more sophisticated through AI and machine learning, but currently it is relatively easily to anticipate and prepare an appropriate musical response for every possible outcome that is significant enough to warrant a musical change.

Since the birth of the industry, the toolbox of interactive music techniques has been constantly expanding, and it looks like an exciting journey ahead. The interactive demands of game music system usually increase exponentially with the amount of freedom a player is given within a game. There is an abundance of new music engines with smart acronyms that claim to provide an innovative way of making a hyper-adaptive music system and often game composers obsess with the sophistication and complexity of interactive systems. A word of caution from someone who has written his PhD on interactive recombinant techniques; game music is more than that. Discovering new and innovative interactive techniques is a fun and fundamental part of game music but complex music and game interaction are not always required – some of the best game soundtracks have been surprisingly simple with their use of interactivity.

In my opinion, a successful interactive music system is one that delivers the right music for each moment in the game in an undetectable manner. Players are rarely aware of how clever the music system is, or how many computations it is capable of as what matters most is that the linear version of the music that they are listening to is entertaining and relevant to their experience. Fans might remember a game because of how much they loved its main theme, but it is rare to hear anyone raving about how smoothly the music adapts to various gameplay transitions. In a way, interactive music techniques could be compared to the work of dialogue editors. Their work is crucial for allowing a clear understanding of the story, but it is usually only noticed if something has gone wrong with the speech editing!

How does audio implementation work?

This is a complex field that cannot be covered comprehensively here but my goal is to give you a glimpse of the basics from a composers perspective. Game audio implementation can be an intimidating and challenging field for many composers as it requires a special set of new skills to execute that are not used in other mediums (ex: film music). Some game composers are directly involved with the implementation process of their music while others rely on audio programmers to handle the technical implementation aspects. You can separate the implementation of an interactive music system into two stages: the conceptual design, in which you set the rules that govern the music interaction with the game world; and the technical implementation, in which you practically realize the algorithm within an appropriate software environment.

Conceptual design

This aspect of interactive music is usually straightforward for new composers to understand if they are familiar with how games function. An easy way of thinking about it is to establish conditional statements that specify how different gameplay events will affect musical change. These can be done in a "**when/if X happens then do Y**" format and should be as specific as possible. For example, some basic hypothetical music instructions can be:

- When "the game begins" then "play Level music on an infinite loop".
- If "the player enters the main room" then "stop the Level music and play the Battle music with a 2 second crossfade".

Interactive music instructions do not have to be limited to entire compositions; they can also be used to affect parts of it. For example:

- **IF** "the player health goes below 50%" **THEN** "add a low pass filter to the guitars with a cutoff on 5 kHz and a slope of 12 dB per octave".
- **IF** "there are no enemies on a radius of 100 m" **THEN** "fade out the drum layer over 5 seconds".

You should aim to establish rules for every gameplay outcome that needs an appropriate musical response. If the rules are quite simple, you can just write them down and communicate them to whomever is implementing your music. However, if the system is more complex, you can use flowcharts to visualize them more effectively (for more info see Chapters 2 and 14).

Technical Implementation Stage

After the music is composed and the interactive rules that control its relationship with the game have been decided upon and established, the audio files need to be implemented within the game. Music implementation can be done directly into a game engine such as *Unreal Engine 5*, or by using third-party audio middleware such as *Wwise* that communicates audio information to a game engine. Teaching you how implementation works on a technical level requires an entire book of its own dedicated to each different software environment. To give you a quick taster of how the process looks in a major game engine, here is how you would implement two of the conditional rules from earlier using *Unreal Engine*, one of the most popular and impressive 3D engines that many game studios use.

MUSIC INSTRUCTION 1

When "the game begins" then "play level music on an infinite loop".

To implement this condition all we need is just two objects: an **Event** named "Begin Play" that notifies when the level has launched, followed by a **Play Sound at 2D** audio command targeted at Level music. As you can notice from Figure 0.2, we do not have to code this in C++, we can use a visual scripting system instead called **Blueprints** that can be considerably easier to understand for non-programmers. To loop the Level music, we can simply double click the audio file within Unreal's content browser and check the Looping tick box in the file properties. Our level will now launch the Level music theme at the beginning of the game and loop it indefinitely.

MUSIC INSTRUCTION 2

If "the player enters the main room" then "stop the level music and play the battle music with a 2 second crossfade".

This condition is slightly more complex to implement but again all the programming can happen within the Level Blueprint as well as the 3D environment of the level design. First, we need to setup some method of knowing when the player has entered the main room. To do so we can setup a **location trigger** that is placed within the level and acts as an invisible

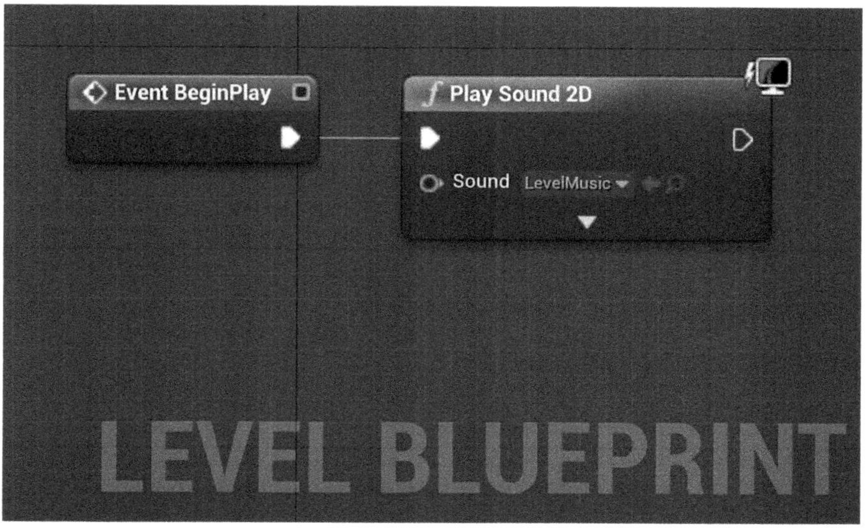

Figure 0.2 A screenshot of a UE4 Blueprint demonstrating how to program the game to play music when it begins.

barrier (Figure 0.3). As soon as the player goes through the box it will "trigger" any musical change we assign to it. All we need to do is right click the trigger and add an event in the level Blueprint that connects it to a **Fade In**, and a **Fade Out** audio command; each targeted at the appropriate music and set to a fade duration of two seconds (Figure 0.4). We now have a functioning interactive transition between Level music and Battle music that is activated every time the player passes through our invisible location trigger.

Location triggers and **gameplay events** in combination with simple audio commands such as **Play Sound** and **Fade in/out** are some basic tools that can unlock a plethora of different interactive possibilities and exist in every implementation environment. Once you feel comfortable with these basics you can also introduce musical change by using continuous **variables** that track a range of game values such as player health, enemy proximity, or any other real-time parameters. It is possible to program much more sophisticated systems within UE5 that take into consideration musical rules such as tempo, meter, or even harmony when transitioning or altering your music. However, because such systems would require a much higher level of programming skill to be executed within a game engine, many composers and sound designers prefer to make use of audio middleware software such as *Wwise* or *Fmod*. Although the learning curve of using this type of software can also be steep for a beginner, it is easier to program complex music behaviours

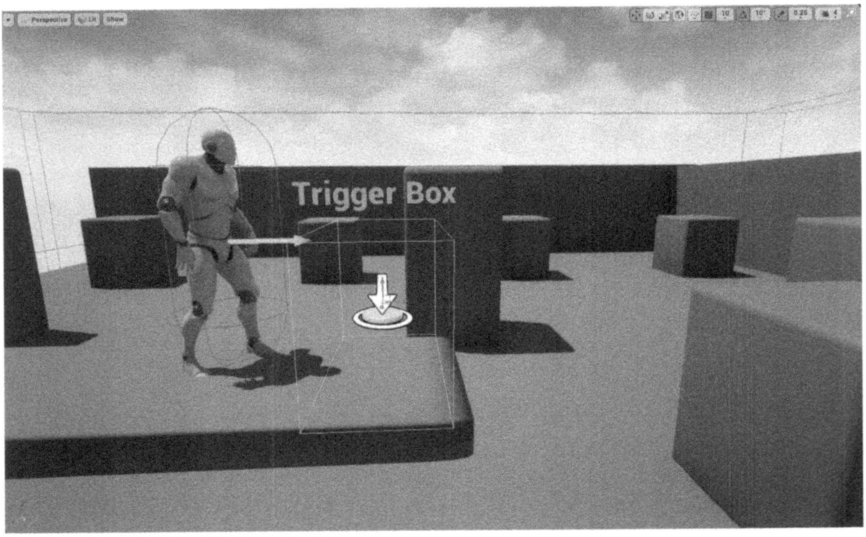

Figure 0.3 A screenshot from UE4 showing how to set up a trigger box inside a 3D game environment.

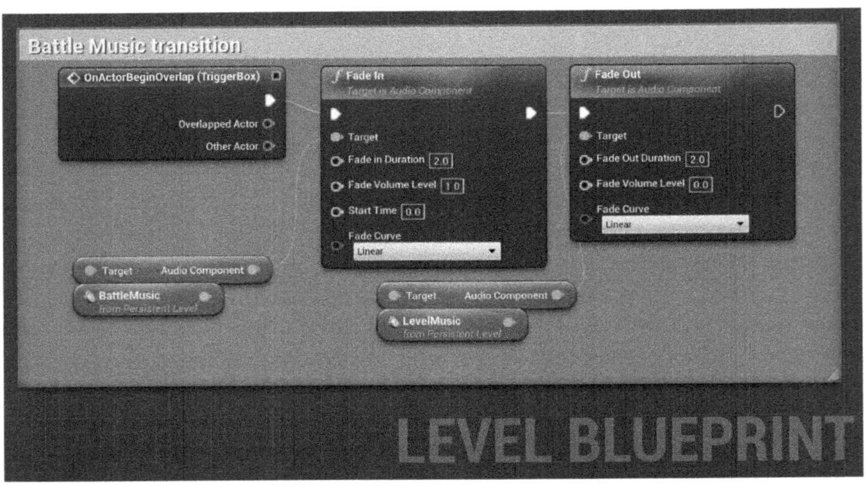

Figure 0.4 A screenshot of a UE4 Blueprint demonstrating how to program a crossfade transition between the battle music and level music tracks once the player activates the trigger box.

natively than within a game engine. This is due to the fact that they are built especially with audio implementation in mind and have a more user-friendly menu-based UI that is closer to that of a traditional DAW.

Career

How can I break into game composing?

This is by far the most common question I receive from composers, and rightly so, as it can be very difficult to get your foot in the gaming industry if you do not know where to start. Getting composing jobs is not a straightforward process and at times you need to apply creative solutions in your approach. Make sure to read the "**How did the composer get the gig?**" section in each of the game studies in the book for further inspiration as well as the career tips from industry experts in Chapters 15, 17, and 19. Aside from the obvious necessary steps of building an online presence in social media, having a professional website, and having a great portfolio on ready to be sent, here are some other useful ideas.

Game jams

One type of event that is unique to the game industry and can be a fantastic starting point for new composers are game jams. These are short competitions in which people from all areas of game development meet up and form small teams with the goal of producing a working game prototype. The design concept is revealed only at the start of each jam, and the participants usually only have a few days to complete it. There are hundreds if not thousands of game jams happening all over the world, some entirely online, and others within a game studio or venue that hosts them. There is a huge variety in the rules of each game jam in terms of the genre, size of the team, duration, and awards of the winning games.

Unless you manage to win some big award, game jams are obviously unpaid, so this is not a way of making money. However, it is a fantastic opportunity to meet other people in the industry as well as build your portfolio with original work rather than rescores of other games. It is important to choose your team mates carefully as it can be a hit or miss experience with some jams being absolute time wasters and others leading to a satisfying outcome. You must be able to work very fast and ideally implement your own work to get the most out of it. It is not uncommon for games developed in a jam to carry on with development and eventually mature into a finished indie project. The Steam and Nintendo Switch game *Nuclear Blaze* that I designed the sound effects for started as a game jam project in the infamous Ludum Dare jam.

Mods

Game modifications (or mods) are another unpaid option of getting your foot in the industry that can be educational and a lot of fun if you work on the right project. Mods are based on modification of pre-existing games and can be made by one fan or very large teams of hundreds of enthusiasts, and can range from slight alterations to complete overhauls that look like an entirely new game. Successful mods can be extremely popular among players and is a great way of gaining exposure and experience. For example, when I contributed some original music to *The Third Age* mod that recreated the world of Tolkien's middle-earth based on the *Total War* engine, the mod reached over 10 million downloads! Other mods such as the *Defense of the Ancients* (*DotA*) that was based on *Warcraft III* became so popular that produced its own multi-million dollar franchise and gaming subgenre. Have a look at Moddb.com to see some examples of what the modding community is currently up to.

Work as a composer's assistant

Many game and film composers often require assistance with their work, and this can be a good opportunity to gain some experience as well as some income. These can be one off project-based gigs or more permanent arrangements depending on the needs of the composer. Sometimes these positions open online and if you actively look you will occasionally find adverts by composers looking for a full-time assistant. However, it is more common for composers to hire assistants that they already know and can trust or to ask for recommendations from their network. Depending on the project, assistant duties might range from MIDI programming, transcriptions, Sibelius edits, and occasionally additional music. Being reliable, resourceful, and easy to work with are useful skills to have as an assistant.

Take the in-house route

If you are also interested in other parts of game audio production such as sound design and implementation, this can be a great option for you. Keep in mind that in-house roles that exclusively focus on composition are quite rare and most game studios tend to work with freelance composers. An exception to this rule are Japanese studios that have large composing teams that work in-house, often for their entire career (see Chapter 13: *Mario Kart 8*).

Network both in person and online

If I am honest, I do not think I have ever met a composer who has claimed they enjoy networking. However, networking can be tremendously helpful with getting more work, but you must also use some strategy and creativity here. There is little use going to expensive game conferences and just handing out your card to 100 people that you speak with for two minutes to give them a generic sales speech. My advice is that networking can be more enjoyable and more productive if you try to make real connections and have real conversations with people you genuinely like and find their work artistically interesting rather than trying to sell your services to anyone who will listen. It is useful to also link up on social media with anyone you meet in person, so it is easy to follow what they are doing and to promote your work in a non-direct way (ex: through general posts in your wall rather than DM spamming!).

You are also much more likely to achieve results by making some research and building connections that seem like a good match for both parties. For example, you can find a small game studio that is based on your local area and then approach the audio director to ask if they have any internship opportunities. Or you can research upcoming indie projects that do not have a composer yet and reach out to them to express your interest by offering a specific vision of how your work would enhance their game using a **relevant** music demo. There are plenty of places to search for upcoming projects, you can start with having a look at indiedb.com but keep in mind that any project that looks quite polished is likely to already have one or more composers on board.

Release and promote music and let developers find you

In the age of the internet, it is not difficult for the right music to reach the right people. If you have something interesting to offer and you promote it well, you might be surprised to find work opportunities coming to you! Here is how this strategy worked out for me: I wrote some music inspired by ancient Greece for a *Total War* mod and placed it on IndieDb.com as a free download for noncommercial use only. This led Canadian developer *Alientrap Games* to discover the music and contact me to write music for their game, *Apotheon* (see Chapter 14). Eventually, I run an advert promoting the links of the *Apotheon* soundtrack with the tags "epic ancient Greek music" on YouTube which was then found by a choreographer that wanted to use the music for the Special Olympics Flame ceremony! Similarly, but in a much much more impressive scale, Gustavo Santaolalla released a personal album of his beautiful collection of compositions for Charango and that led to licensing requests directly from Hollywood directors and AAA game titles (see Chapter 11).

Promoting your music effectively needs a book of its own but one tip I can give you is to look at digital distribution services such as CD Baby that can upload your music to every platform simultaneously and track sales for a small fee (ex: £50). Another tip is to upload your music directly to game engine asset stores that game developers might use (ex: Unity or UE5), as well as to reputable production libraries online (ex: Audio Network).

Is working in game audio a viable career?

It depends on your individual expectations and abilities. This industry can be emotionally rewarding as well as financially lucrative for successful composers, and game music is increasingly popular even outside games. Over the last two decades the industry has witnessed impressive growth, and video games are now the biggest form of entertainment globally with current revenues surpassing the film and music industry combined![4] Such unprecedented growth naturally creates a massive demand for new music, but unfortunately, it is incorrect to assume that this is all good news for composers – the demand for game music has risen exponentially but so has the supply of composers. There is not enough data to accurately quantify the available opportunities to the ratio of game composers but my own experience in the industry has shown me that in many countries the pool of aspiring composers far outstrips job availability – with the most desirable gaming projects naturally being the most competitive.

If you attend any game conference in a big game development city (ex: London, or Montreal), you might find an overwhelming number of aspiring composers waiting in line to speak to game developers. According to anecdotal conversations I have had with game developers, most unsolicited applications they receive are from composers, which makes it by far the most competitive profession in the gaming industry, closely followed by 3D artists. If you examine the posts in almost any game development forum or social media group, you will quickly be disappointed to find that composers are dominating the "looking for work" section by a big factor, and sadly, are often spamming multiple unrelated channels with links to their music in a desperate attempt of self-promotion. I honestly believe that the number of aspiring game composers online now might be comparable to the number of aspiring Hollywood actors in L.A! It is only natural for people to want to work in something that is so much fun and that they are deeply passionate about.

However, not everything is doom and gloom. If we account for all the different platforms ranging from PC, consoles, mobile, and other emerging mediums such as cloud gaming and VR, there are tens of thousands of new games released every year and that number appears to keep growing. In 2021 on the *Steam* platform alone, there were approximately 30 new games released every single day while at the time of writing there were approximately 477,877 mobile games available to play on *Google Play.* Considering the sheer number of games in development, it is very likely that with some skill and patience you will find composing opportunities if you actively pursue them. I genuinely believe that each composer has something unique to offer and often it is about finding the right project and the right team of collaborators that matches their skills and style.

Countries across the world such as the USA, the UK, Canada, China, India, and Japan have a strong game development job market that frequently seek numerous full-time in-house game audio roles. Obviously, the availability of these openings is subject to different working visas and language requirements, and it is not easy to gain employment sponsorship for work outside the country you live in, except in rare circumstances. You can go to gamedevmap.com that lists every game studio in the world with over five employees to see which game studios are situated in your local area. For example, in 2022 there are over 574 registered game studio entries in England, and new audio design positions open on a regular basis through various job boards (ex: uk.Indeed.com).

In terms of financial expectations see the questions on the following pages but unless you manage to score the next AAA franchise you should expect that it might take time to grow your business to the point that you have a steady flow of income you can rely upon, and you should expect that this figure can fluctuate considerably over time. Think of a game-scoring career like a marathon rather than a sprint and strategize accordingly by being flexible and entrepreneurial when needed. The good news is that most of the skills you need in the gaming industry are transferable to other media. I personally truly enjoy moving between industries depending on work opportunities and frequently work in theatre, art exhibitions, animation, and films.

What are some alternative career paths in game audio?

If you are passionate about game audio there are also other roles in the gaming industry that can be complementary to composition. Here are some of the main job descriptions and approximate annual salary expectations based on a number of available job posts in the UK during 2023.

Audio designer

Audio designer positions are usually a jack of all audio trades with a primary focus on creation and implementation of SFX using game engines and middleware. The role might also involve other tasks such as recording voice overs, mixing audio for cinematics, and occasionally helping with composing music. It is a competitive and highly skilled position, and depending on the size of the studio there might be several audio designers within the audio team. As an example, a mid-size company could have a team of up to ten full-time audio designers working across multiple projects. A starting UK salary would be approximately £32k for entry-level positions, while senior roles currently advertised are approximately around £46k+.

Audio programmer

This is very different than your usual audio position as it requires strong programming skills first (usually in C++), and general audio skills second. Audio programming roles are currently some of the most in demand technical positions in the field and have great prospects in terms of salary as well as career development due to lower competition. You can expect to make upwards of £40k+ depending on experience.

Audio engineer

This is your typical audio engineer position and is usually centered around recording and editing dialogue and occasionally field recordings for any other audio assets as needed. Unfortunately, is not usually well paid unless you have demonstratable experience in multiple AAA projects.

Audio director

Each audio team is led by an audio director which is typically the audio designer with the most experience and seniority. This is a managerial position in which you are the main point of contact with the heads of other departments, and you are responsible for shaping the audio vision for one or more games that are in production as well as hiring external composers. This can be a higher salary position exceeding £55k+ in the UK, and going much higher for major studios especially in the USA, but it requires extensive experience within the industry (usually 5–10 years). Many audio directors first begin as audio designers and then move up the career ladder.

Orchestrator

Larger projects that can afford to record real orchestras often rely on the help of freelance orchestrators. To better understand this role, I have interviewed composer and orchestrator **Lorenzo Bassignani** who has worked on Sony's *Horizon Forbidden West* (2022).

Mini interview with orchestrator Lorenzo Bassignani (Horizon Forbidden West)

MA: What does an orchestrator do?

LB: In the world of music for media basically an orchestrator brings the music from a digital medium (a MIDI mock-up) to a real one (in this case the orchestra) while making sure that what is written works for each instrument (ex: by balancing the instrumental sections, editing the dynamics, and sorting out all the articulations). The role often includes preparing the parts and the conductor score. It is a very challenging job as not only you need to make sure what you put on paper works but also you need to make sure you interpret what the composer wants correctly.

I work for composer Joris de Man, who is one of the most in-demand composers in the videogame industry. For Sony's *Horizon Forbidden West*, I provided extra support towards the end of the studio recordings, and my role was as a score copyist/orchestrator, working together with Joris's assistant and the conductor. After that, I assisted on Warner Brothers' *Gotham Nights* and other games for Joris and his team, both as score copyist/orchestrator and also as digital orchestrator (orchestrating using samples).

MA: How did you get the job?

LB: It was through an advert. The ad was for an assistant. The applicants were narrowed down and I got to the interview stage. I'm told it was a close call and it was hard to choose. I did not get the job but a few months after, I received a call from Joris's producer, as some extra people were needed. That is how I started working for him on a freelance basis.

MA: What did you find the most challenging part of the job?

LB: Every step along the way is hard and requires a lot of concentration: in addition, it is the speed required for each job that makes it even harder. There is also always something to learn which demands some extra research (a new instrument or an unfamiliar technique). For instance, certain extended techniques have no standardized notation; this calls for extra elucidation for both conductor and musicians.

MA: Which software do you use?

LB: Not much is required in terms of software and processing power. I use Sibelius together with a plugin for sounds called Note Performer (it's better, smaller, and lighter than Sibelius Sounds). I also recently learnt Dorico, which has the advantage of a real-time function for dealing with individual parts and the conductor score; in Sibelius it needs to be done manually, in Dorico it's possible to expand/condense and deal with the string divisi and the other sections with one simple click. In spite of this, Sibelius is still my first choice, at the moment.

Business and money

What are the main types of copyright deals for composers?

To understand how composition contracts work, you must have a basic understanding of music copyright. There are two types of music copyright that are important to a composer: the **publishing rights** which refer to ownership of the composition, and the **master rights** which refer to ownership of the recording. These rights can be shared by one person or by multiple parties such as the composer, a record label, a musician who performed in the album, a music library, a game studio, etc. If you compose and produce a piece of music by yourself in your DAW, then you obviously own 100% of all the copyright (100% of the Publishing and 100% of the Master rights) unless you decide to transfer these rights to someone else. Copyright is important because it is directly involved in the two main types of deals that typically occur between a game company and a composer – **buyouts** and **licensing**.

Buyout/work for hire

In a buyout deal, all the music along with all copyrights (both for the composition and the recording) are transferred in full to the game studio. This is standard procedure for original game music in most bigger projects with 98% of all AAA contracts being buyout deals and only 2% being licensed.[5] After a buyout is completed the game studio fully owns the music, in the same way they do with other parts of the game (ex: 3D models, programming, story, level designs), and it can do whatever it wants with it, including using it in other projects, organizing concerts, or even selling it to other developers without ever needing to ask further permission from the composers or further compensate them in any way. Salaried in-house composers and game company employees also fall under this category unless they have a special clause in their contracts that specifies otherwise.

You should only do a buyout deal for a game if you think that the payment is worth the price for completely handing over the publishing and master rights of your music. The music will not belong to you, technically the company does not even have to credit you, and you cannot even use the music in your own projects anymore. Likewise, if you hire any musicians to work on your project, make sure they sign a work for hire contract unless you wish to share a percentage of the music ownership with them.

Licencing

Instead of transferring full ownership of the music to a game company, composers can also allow the company to licence their music for a specific use in a game, trailer, concert, or other events. The composer retains all copyright, so she/he is allowed to sell the music, perform concerts, or use it again in other projects unless there are other exclusive license restrictions specified during the contract (ex: the music cannot be licensed to another game for the next five years). This is more common in indie games that have small budgets and cannot afford to do buyouts; according to the GameSoundCon survey 53% of paid indie contracts were buyouts and 47% were licenses in 2021.[6] In some cases, AAA games also prefer to license music if it does not make financial sense to buy it. For example, the soundtrack of the sports game *NBA 2K* includes some of the biggest pop hits from artists such as Daft Punk and Kendrick Lamar. Obviously, it would be insanely expensive and unprofitable for the studio to do an outright buyout of the full copyrights of these hit songs just to use them in one game, but a licencing deal allows them to use it in exchange for a one-off fee that is split between the owners of the Publishing rights (usually the artist) and the Master rights (usually the label). This is also quite common in the film industry and comes under the "synchronization" license.

Shared copyright deals

Technically it is possible to do a hybrid deal that splits the publishing and recording copyrights of the music between the game studio and the composer in an agreed percentage. This would make both parties co-owners of the music and would both benefit of any further commercial uses of the soundtrack (ex: radio/Spotify royalties of the soundtrack). However, this approach creates the need for additional management and communication and would require copyright clearance from both parties for any future uses of the music outside the project. Using special clauses in a contract (ex: regarding soundtrack sales) might be simpler than doing a shared copyright deal.

Keep in mind that most indie developers do not really understand music copyright, so if there is no transfer of copyrights mentioned in a contract (or in the email agreements) then you still maintain all rights to your music, and in effect you are doing a licensing deal where you allow them to use your music in their game.

How much money do game composers make?

As you would expect, composer fees can vary enormously within the industry depending on the individual composer as well as the size of the game project. These fees are usually negotiated on a "per minute of completed music" basis that is multiplied by the number of total minutes produced. For example, in a hypothetical agreed rate of $100 per minute, a three-minute composition will be sold for $300 for a full buyout, and a one-hour soundtrack will be sold for $6,000 (60 minutes × $100). One of the most reliable sources of data we can examine to get a general overview of these per minute salary ranges is the *GameSoundCon* survey which is conducted annually and includes data from over 600 game composers from around the world.

According to the 2021 GameSoundCon report, the composer fees reported on indie games varied dramatically from $100 per minute all the way to $1,500 per minute of music. Such wide differences can be attributed to the fact that the production quality and success of indie games also vary enormously, from games that are played only by a few friends and family members to games such as *Minecraft* that was started by one person before being bought by Microsoft and becoming one of the best-selling games of all times. From my personal experience of negotiating fees in the indie world for over a decade, I would say that a $200–400 per minute range is a realistic number for serious indie games, and only a few development teams would outright pay over the $500 per minute mark unless the project gets sufficient traction on Kickstarter or other forms of social funding. To be completely transparent, I have also frequently encountered projects that offered even less than $100 per minute, projects that only paid in shares of (potential) game sales that might never come, and of course, the occasional obnoxious developers who would ask people to work entirely for free so they can gain "experience". If you ever decide to work without payment, at least make sure you do a sync deal instead of a buyout. It is noteworthy that these types of deals are possibly underrepresented in industry surveys because composers who are working for free might be less likely to report it.

According to the same survey data, the most common fee for a composer working on a professional mid-sized project was approximately $300–$500 (USD) per minute of completed music. This means that a game theme of about 3:30 minutes sold for $1050–$1750, and a game soundtrack of 60 minutes generated between $18k and 30k for the composer. Finally, on the AAA spectrum the most common budget was $1,000 per minute, or $60k for one hour of music, but some composers received much higher fees surpassing the $3k per minute and $180k per hour to even $5k per minute and $300k per hour of music for the top game composers. Keep in mind that the total amount of music required also varies and can be much longer than 60 minutes of music; a major game like *World of Warcraft* had over 45 hours of original music composed by a team of composers over multiple years, so the budget of the music was likely on the multi-million-dollar range. There is no official data for Hollywood celebrity composers working in games such as Hans Zimmer or Gustavo Santaolalla but considering that the biggest AAA games have budgets over the $100 million mark that equal or surpass Hollywood blockbusters, it is safe to say they can afford to pay astronomical fees for music if they believe that a composer will genuinely add value to their product.

Alternate forms of payment

Payment for writing game music, especially for smaller or mid-budget projects, does not necessarily have to be done 100% in cash, and there are other creative ways of negotiating payment that can be beneficial to both parties. One common example is a hybrid deal that would include some payment in cash, ownership of a portion of the game through royalties, and a percentage of soundtrack sales in Steam (usually 50%). As you can see in Figure 0.5 indie developers with smaller budgets often offer a lot of non-cash incentives, but AAA projects prefer to do cash buyouts. This is to be expected as which company would want to give a composer a game royalty percentage from a billion-dollar game franchise?

By taking a per unit royalty in exchange of a smaller cash pay is a way of becoming a business partner and having a direct stake to the success of the game which can obviously go either way. I have worked in games that made little or no money as well as many that did not even get finished. However, when I worked with the game studio *Alientrap* they wanted to make me feel I am part of the team, and that the success of the game should matter to me, so they gave me a tiny share of the game ownership in exchange of lowering my composing fee. This worked out great for everyone as Sony ended up buying the exclusive rights of the game for the PS4 (over the Xbox) for a very rewarding fee!

If you are taking your first steps in game composition and you are working on a very small project that has no budget, you can negotiate to be paid

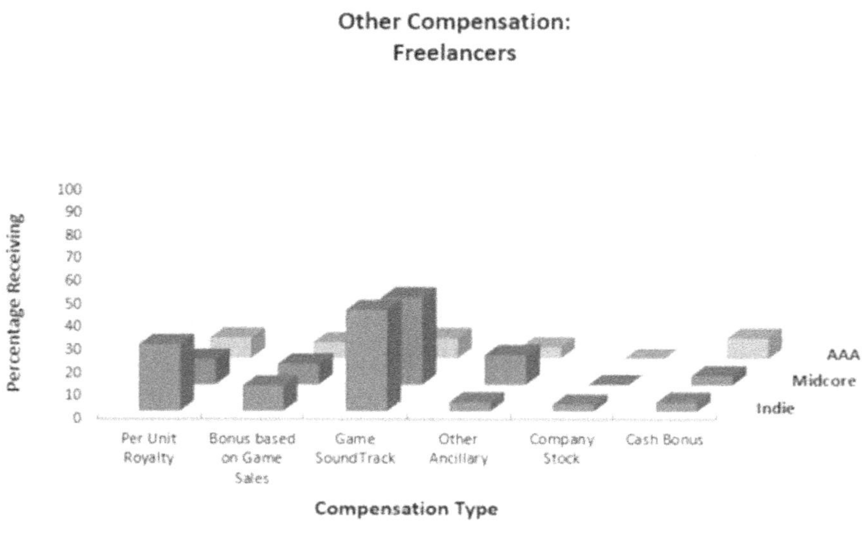

Figure 0.5 A chart from *GameSoundCon* showcasing the type and percentage of alternative forms of payment that composers received in different sized game projects.[7]

in shares of the game, and/or do a licensing deal where you allow them to use your music, but you retain the copyright. Indie game composers also have more flexibility in negotiating alternate forms of payment which can also push their income to higher levels, especially if the game or the soundtracks sell well. These additional sources have been baked into the reported income of indies reporting $1k+ per minute range which is entirely possible after the games become successful.

In-house composer salaries

The nature of in-house composing work can be quite different from the risks of the freelance world as you are paid a yearly salary, and your job is not as directly impacted by the amount of music you produce or how well the game sells. In-house positions that are exclusively focused on composition are very rare, and over 80% of in-house composers from the GameSoundCon survey reported that their job role required them to work in additional areas of game audio such as implementation, sound design, or audio engineering. It is therefore quite difficult to distinguish the exact amount of payment reserved for composition. Looking at salaries of in-house game audio employees that have roles including composition, the reported numbers are:

1) for the USA $115k average yearly salary, $92k median yearly salary,
2) for the UK/Europe £43k average, £36k median salary (in British Pounds), and
3) for the rest of the world $55k average yearly salary, $49k median yearly salary.

These numbers demonstrate a much higher income for USA-based jobs, but it is worth considering the significant disparities in the cost of living between working in a game studio in San Francisco and a studio in Warsaw Poland (ex: CD Project RED) which could easily be three times higher.

Will buying expensive gear improve my sound? Does my studio need to look like a spaceship?

The appeal of a shiny piece of new gear (or an old rusty piece of vintage gear if that is more your thing), commonly known as gear lust, is certainly understandable and affects all of us. Marketing campaigns try to convince us that we need to spend a lot of money in expensive audio gear to achieve a professional standard but do not fall into the trap that this is a requirement. I have occasionally witnessed students feeling discouraged after looking at dream studios of other media composers such as the one owned by Junkie XL (video example 2). Junkie XL acquired his outstanding gear over decades by patiently buying used equipment after they fell out of fashion. Interestingly, according to his social media he has recently decided to sell a big part of his collection as he found himself underutilizing most of it, and to downgrade to a smaller minimal setup consisting only of his favourite gear.

The fact is that your skills affect the production standard of your music much more than the price of your studio as expensive gear alone will have little impact on the quality of your work. I suggest researching how multiple award-winning musicians have produced major hits by relying only on stock plug-ins. In my experience of teaching production students, far too much time is devoted in researching and buying gear rather than learning how to use it. Chances are that your current DAW has more power than you realize and has everything you need to achieve a professional sound, at least regarding the fundamentals of your mix. It might not have a shiny UI compared to the latest plug-in bundle, or it might not have the best pre-sets ready to go, but it will achieve a similar result if you spend the time learning how to use it skillfully.

My advice here is not to avoid buying audio gear all-together and become an audio minimalist but just to pause and consider how much of a difference a purchase will actually have on your sound before making an investment. Will your guitar sound better if it is recorded through a *Neve* mixing desk that costs £200.000 over a *Focusrite Scarlet solo* audio interface that costs £200? Absolutely! But if you listen to a blind test of both recordings, and you cannot tell them apart then maybe you should save your money and use it in something that will have a genuine impact. If you cannot test gear physically in a local dealer, there are plenty of YouTube blind tests of any audio gear imaginable that are good enough despite the YT compression, so use them and make your own informed decisions of what you personally prefer.

However, buying new gear can sometimes benefit your music in another way that in my opinion might be more important – inspiration! After relying on the same setup for a while, you probably develop certain workflow habits. Introducing a new instrument or FX that you are not familiar with might take you towards a different creative path that you have not considered or explored in the past. When I bought my first *Moog Grandmother* I somehow found myself making monophonic arpeggiated square wave sequences for hours. I might have been able to get a similar and more precise sound much faster by using a virtual plug-in and a mouse, but it wouldn't be as much fun or as inspiring as using a hands-on analogue device, and it is hard to put a price on that!

To get a better idea of some of the gear that professional game composers have used, make sure to read the **Production Tools** section of each game in the book. As you will see, some of it might be exotic and expensive but other tools might be widely available and affordable, and it is the creativity behind their use that gives them its characteristic tone. In terms of audio implementation software, the good news is that as of the time I am writing this book, all major software and popular game engines can be downloaded free of charge if they are used for educational purposes only. If you have enough hard drive space you can download *Unreal Engine* and Unity, as well as the industry standards of middleware *Wwise* and *Fmod* completely free of charge. Some of these software would normally be very expensive, but they have changed their business model to a royalty system in which you pay a percentage based on the sales of a game (usually 10% for games that generate over 1 million dollars in profit). The learning curve of these software can be significantly steep, so I strongly suggest you start with only one of them.

Education

Should I go to college/university if I want to pursue a career in game audio?

Deciding to pursue a formal education in music has both pros and cons, and the education quality and career prospects you will receive from your studies can vary significantly between different programs. After having spent over 15 years both as a music student and teacher across seven different universities in the UK and the USA, I believe I have an excellent insight to help you with this decision. The following are the major factors you need to examine when considering a particular program.

Career prospects and certificate value

Obtaining a formal qualification can be beneficial to your game audio career but that also depends on the type of work you pursue. For freelance composition jobs, qualifications do not make a big difference in my opinion but if you have graduated from a prestigious university, it can help build some credibility. However, for full-time positions having a degree is becoming more standardized. Looking at recent trends in game audio job postings there is a clear trend that shows that game studios have started to prioritize formal qualifications in their job requirements: over 90% of recent hires in game studios have a formal education, and over 90% of new game audio professionals have a formal qualification in a related field, usually in music or production (Figure 0.6). However, having years of experience in the industry is a more significant factor than education in terms of salary and desirability to employers which indicates that if you can manage to get your foot into the industry and score some recognizable credits then you can have a successful career regardless of your prior education. Having a degree also allows you to pursue other music related jobs that can be complementary to your freelance composition career, such as teaching music part-time.

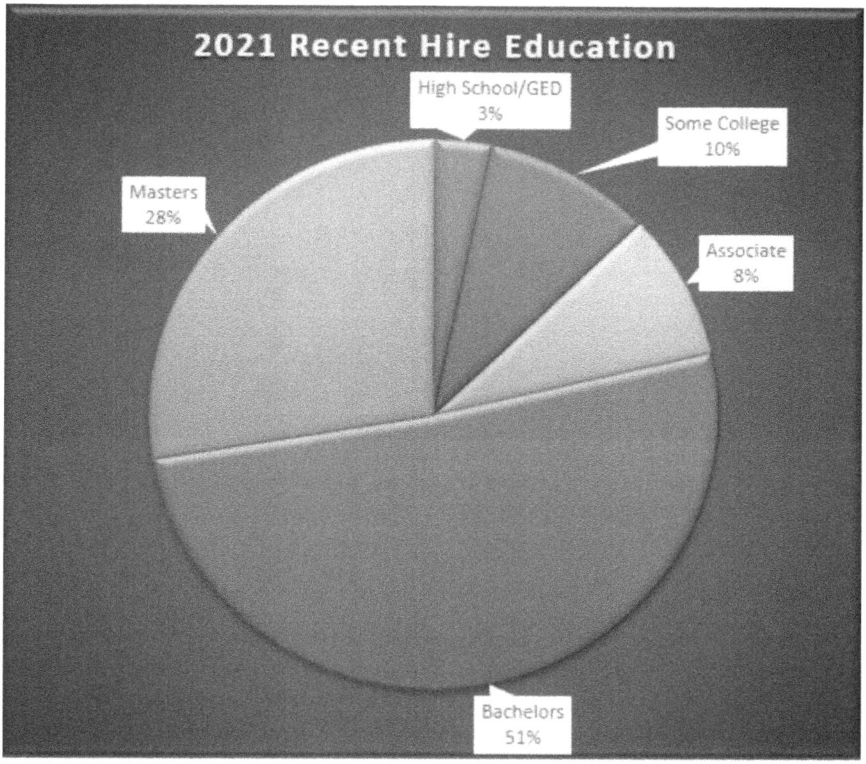

Figure 0.6 A chart from *GameSoundCon* showcasing the education qualifications of new game audio salaried employees with two years or less experience in the industry.[8]

My advice is to look at the graduate data from the career services of the university you are considering and to inquire further about the methodology used to collect it. Often institutions claim that most of their students find employment soon after graduation, but this information can potentially be misleading because their career services might consider a student as employed on the merits of releasing their own album on Spotify. As most musicians make a living from freelancing it might be more effective to examine recent graduate income/salary. Check what are the work and creative opportunities that the school will offer you both during your studies as well as any internships connections they might have for after you graduate. For example, the Game Audio program I have designed for GSMD in London offers students the opportunity to record with a real orchestra in a world class performance venue (Figure 0.7).

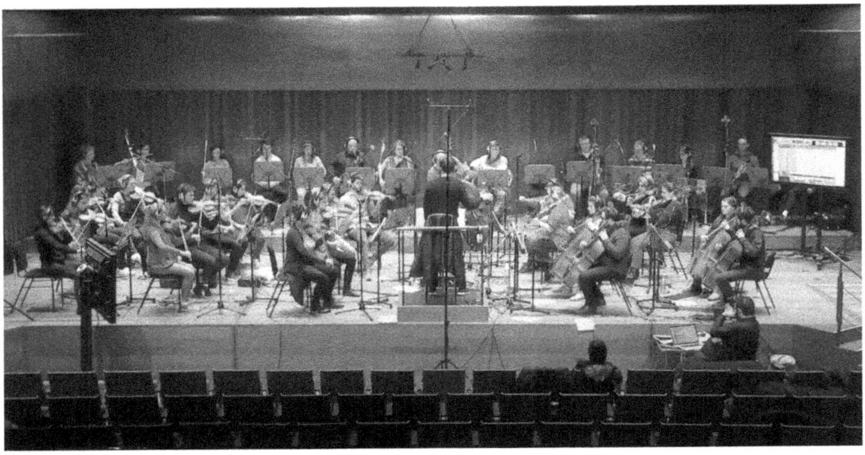

Figure 0.7 A photo taken from a GSMD game audio recording session with the Guildhall Session Orchestra in Milton Court.

Financial costs

Tuition fees for a three-year degree can range from completely free in some state funded EU universities, to over $160K for prestigious USA institutions such as NYU or USC. On top of these tuition fees, you must anticipate the cost of living in major cities such as New York or London that can be substantial as not all universities offer student housing for the entirety of their programs. It is entirely possible for a student studying a four-year music degree in New York City to need $250K to fund her entire studies, a cost that will be extremely challenging to earn back for most graduates. Education in the UK is more affordable than the USA, and most universities cost about £10K per year for UK students and £17K for international students in 2023. There are many options across the EU both in private English-speaking programs and public institutions usually in the national language of each country.

Time commitment

This can vary from 1 year certificate programs, 1–2 years for Masters, 3–4 years for bachelor's degrees, and 3–5+ years for PhD. Many programs are labelled as full time, but this rarely adds up to 40 hours of classes each week. Most likely it is a combination of lectures, workshops, optional events, individual study, and homework. It is worth checking the specific time commitments of the program you are looking for especially if you are considering working part time during your studies.

Curriculum

The curriculum might vary considerably between programs and institutions. Look at the syllabi, curriculum, and assignments of each year of your studies to get a better understanding on what you will be learning and how it aligns with your current knowledge and goals. When I studied for my bachelor's degree, most of the composition classes available in my college were based almost entirely on 20th-century classical music, and commercial music genres were frequently ignored or even frowned upon by some of my teachers. In recent years, music education institutions are becoming more open minded and the issue of musical elitism appears to be diminishing but it is definitely worth looking into the musical culture and diversity of the program you are considering.

Support

A good institution will offer you support in many different levels: tutoring, financial, career placements, and mental health.

Community/networking

In my opinion, this is one of the main reasons to attend a university in the first place as it is likely that this is where you will develop lifelong friendships and professional connections. Belonging to a music community is a vital part of your studies as being able to share your work, receive feedback, get inspired by the work of others, exchange ideas, and collaborate with your peers is tremendously helpful. I meet some students who argue that if they quit their degree and go at it alone, they will have more time to focus on their music and while that might be true, I find that few of them succeed in doing so, unless they already have a network of musicians in place.

It is worth examining the type of community and culture around the institution you are considering and making sure this is a good match for you. As an example, when I started my music studies in the UK countryside for my A levels, I was the only international student out of 500 English pupils which I found a challenging environment to intergrade socially as a 17-year-old Greek student. Finally, be mindful of any programs that are delivered entirely online as they might be lacking this communal aspect all together.

Access to production facilities

If you are paying high tuition fees you should expect to get access to high-end production facilities and recording studios. Make sure to check how much time share each student is allocated during their studies as often these facilities can be overbooked and overwhelmed thus being almost inaccessible to new students.

Should I focus on developing a niche or try to write in many styles?

Both approaches are valid as each has their own advantages and disadvantages. Being a stylistically diverse composer can be great fun and allows you to pitch your music to a wider range of freelance projects. However, promoting yourself as an all-around composer might be challenging as it might appear inauthentic. In today's internet age that game studios have access to a giant pool of potential collaborators it is only reasonable to expect that they will prefer composers whose previous work seems like a genuine match to the specific style of music they are looking for. For example, if a game in development is looking for a dubstep-based soundtrack, a producer who is fully immersed in that scene might be more appealing over an orchestral composer who occasionally writes dubstep on the side along many other genres. An exception to this rule would be when studios are looking to fill full-time inhouse positions, as composers who can write well in multiple styles at a good standard will have an advantage as they can contribute to multiple projects of the studio as needed.

Marketing issues aside, the main challenge for an all-around composer is the following: can you actually write and produce great music across multiple genres at a standard that can beat your competition? If you can, then go for it and just make sure to tailor your application for any pitch with what is most appropriate for each individual project. However, I have met many students that think they can quickly learn to write in any genre: How hard would it be to learn some jazz? Just add 7ths! Traditional Indian music? Just use a Santoor plug-in! Produce a Synthwave hit? Just use an analogue synth with an arpeggiator! While such superficial approaches might be passable at an amateur entry level, writing and producing music in an unfamiliar style requires a deeper understanding which will take time and effort to develop.

Developing your own compositional voice and sound can certainly give you the advantage of being different from every other game composer. I have personally been given many opportunities because of my niche interest of writing music that is influenced by ancient Greek and Roman aesthetics. In my opinion the two drawbacks with having a specialty are (1) that you narrow the scope of suitable projects, and (2) that you might get tired of writing in the same style after a while. However, developing a unique style does not mean that you have to remain static as an artist for the rest of your career, you can still develop your voice and recontextualize it as needed, as well as gradually add new techniques in your toolbox (this is what this book is for!). One liberating way you can use to explore new directions while avoiding any marketing or artistic identity conflicts is to use different artistic aliases.

What skills do I need to have to pursue a career in game composition?

Becoming a successful game composer should be about being a good composer, right? Well, unfortunately it is a bit more complicated than that as writing music for games can require additional skills than composing music for other media. Regardless of how you choose to gain these skills, here is what I believe you need to know to have the best chances for success.

Composing

Well, this one is obvious, the better composer that you are the better your chances of people wanting to work with you! Even though who is a "good" composer is entirely a matter of personal taste that cannot be objectively measured, actively listening to new music, learning new techniques, and practicing your craft on a consistent basis is very likely to increase your chances of writing music that others will find engaging.

Production

This should be obvious, but it is often shockingly underestimated by many composers and educators, especially those who come from a traditional classical education. The production quality of your work is a fundamental factor on how people perceive your music even if they are not aware of it. A strong production can sometimes elevate even a mediocre composition, while a poor production will always diminish even the most evocative piece of music – just imagine listening to your favourite piece with unwanted distortion or strange filtering, the emotional impact of the music can be ruined. You will also need to be able to produce music fast, at a high standard, and relatively cheap, so unless you want to constantly rely on paying sound engineers and other musicians for every part of the process this area should be a key priority.

Implementation/programming

One of the most common questions of new composers entering the game audio industry is if they need to learn how to code. As we will see in many of the case studies in this book, many well-known game composers working in big AAA games were not involved at all in the implementation of their music as this was handled by audio programming teams. In my opinion, you can certainly work as a game composer without having to deal with the technical challenges of implementation beyond a basic understanding of how interactive music and implementation works. However, learning how to do some audio implementation on your own offers many benefits. Firstly, unless you are working with a studio big enough to have a dedicated audio team chances are that the game programmers are going to have their hands full with an endless catalogue of other things to fix; and trust me on this, testing some innovative music system will fall at the far end of their list of programming priorities. I have had professional projects rushed to the market in which the audio was clearly distorting under certain conditions but the pressure to release the game in time meant that there was no time for the programmers to fix this as they simply had bigger problems to worry about. Knowing how to implement your audio files yourself will give you the advantage to instantly test your music in the game engine as you are writing, as well as make tweaks to improve the details.

Secondly, having implementation skills can also open other possibilities such as working as an audio designer in a game studio. Learning how to do your own implementation does not necessarily have to involve learning how to write code but depending on the software a game is built on, it can also be done using visual based programming systems such as Blueprints in *Unreal Engine 5*, or audio middleware software such as *Wwise*. There are also books available that can help you to learn basic implementation skills without ever writing a single line of code even in game engines that are more code reliant such as Unity.[9]

Business and marketing

Freelance work makes up approximately half of all game audio work contracts, so if you take the freelance route you will need to learn how to effectively operate a business.[10] This includes managing your budget, managing your own working hours, keeping accounting books, filling your business taxes, marketing and selling your music, hiring musicians, writing contracts, reaching out to new clients, etc. You can always outsource some of these skills when needed but that will add to your costs.

Communication and collaboration skills

Even if you enjoy working alone, making video games is a collaborative process so this is a job in which a certain amount of teamwork is required and being a good collaborator is highly valued. You will have to learn how to communicate your musical ideas effectively without relying too much on music terminology, as well as exchange constructive feedback with other team members in a respectful and professional manner. If you are difficult to work with few people will want to work with you even if you are a good composer. I have occasionally heard from students the idea that famous musicians are usually self-centered and arrogant so somehow this is a personality trait they should model. What they fail to realize is that obnoxious artists can sometimes get away with unprofessional behaviour because the skills they bring to the table are so high that people might be willing to temporarily tolerate them, but it is certainly not something that helps their career in any way and more often than not can also end it.

Well-being skills

The competitive and uncertain nature of freelance work can often feel like a stressful environment to navigate so it is vital to learn how to take good care of yourself both mentally and physically. I see many students who get obsessed with "making it big fast" and then on the first sign of rejection feel like a "complete failure" and want to give up, but the reality is rarely that black or white. I encourage you to always keep a growth mindset as you might need to adapt your strategy or re-invent your artistic identity multiple times over the duration of your career no matter how much success you achieve. Sometimes you might need to work hard to meet a big deadline, but usually working consistently over time is much more productive and viable than relying on pulling all-nighters in the last minute. Also remember that other life activities like going for a run, doing something fun with friends, or just resting might be just as important than working on your music. Finding a work-life balance that works for you is key to having a sustainable composing career you enjoy!

Takeaway tasks

Task 1 – Analysis (easy) – Analyse the music functions in a game level

Play a game of your choice, with and without its original music, and write a bullet point list describing how your experience is different. You can use the five categories above or you might even discover some additional uses of your own.

Task 2 – Analysis (easy/medium) – Analyse the musical style of a game of your choice

Find a game that you consider as having a musical style that is strongly suited to the experience. Write a few bullet points on what characteristics makes you think so. Aim to expand beyond the obvious by doing some research and analysis. For example, "I think that the music in *Skyrim* sounds epic" is a bit generic, something more specific could be

> I think that layering multiple takes of a massive male choir singing in unison in a made-up language, recorded at large wooden hall with acoustics resembling a Viking Longhouse, and melodies based on a mixture of ancient Greek modes, is effectively evoking the comradery and wild spirit of the Dragonborn culture!.

Takeaway task 3 – Composition (medium) – Write a short theme based on a piece of concept art of your choice

Here are some tips you might find helpful:

1 Spend some time carefully studying the picture, especially look for details that provide clues about the story and/or the required mood. Write a couple of sentences describing your observations. For example, while looking at the concept art from the game *Horizon Zero Dawn* (video example 1, 5:00–5:30) I made the following observations:
   - There is a young female protagonist wearing tribal clothing and holding a bow perhaps suggesting she is a warrior or a hunter.
   - There are colossal dinosaurs that appear to be robotic and technologically advanced.
   - There are some modern city ruins on the edges, possibly indicating some apocalyptic catastrophe has taken place in the past.
   - The environment is lush with wild nature and full of life despite the apocalypse.
   - The sunrays passing through the forest trees, the flying birds above the robots, the snow-capped mountains, and the big open sky perhaps indicate a feeling of hope and adventure.

2 Think about which are the most important elements to you, and how can you represent them in your music. You can begin by building a pallet of suitable instruments that relate to your interpretation (solo voice? tribal percussion? futuristic synths?). Perhaps there is a strange scale you can use, or a production effect that relates to your concept.

3 If you immediately have an instinct on what to do just go for it! If you are still not certain how to start, spend some time listening to reference tracks of similar soundtracks. You do not have to always reinvent the wheel; research on how others have approached similar stories in the past might spur new ideas.

Task 4 – Analysis (medium) – Analyse the implementation of a game of your choice

Can you identify some of the rules that control the behaviour of the music? Try to be as specific as possible with your observations and break them into two parts:

1. What prompts musical change? Look for specific location triggers, game events, and/or continuous gameplay parameters (variables).
2. How does the music change? Look for crossfade transitions, use of transition segments, use of layers, changes in tempo, addition of effects, modulation, or any other type of development that you can notice.

Task 5 – Composition/production (medium) – Participate in a game jam!

Go to https://itch.io/jams to check a detailed calendar of upcoming game jams that happen online. There are also many game jams that happen on location so you can do your own research in your local forums.

Notes

1 Mohn et al., *Perception of Six Basic Emotions in Music.*
2 Aristopoulos, "A Portfolio of Recombinant Compositions for the Videogame Apotheon."
3 Semel and Reznor, "Vintage Interview: Nine Inch Nails' Trent Reznor."
4 "Gaming Worth More Than Video and Music Combined."
5 Schmidt, "Game Audio Industry Survey 2021."
6 Schmidt, "Game Audio Industry Survey 2021."
7 Schmidt, "Game Audio Industry Survey 2021."
8 Schmidt, "Game Audio Industry Survey 2021."
9 Coggan, "Unity Game Audio Implementation a Practical Guide for Beginners."
10 Schmidt, "Game Audio Industry Survey."

Bibliography

Aristopoulos, Marios. "A Portfolio of Recombinant Compositions for the Videogame Apotheon". 2017. https://openaccess.city.ac.uk/id/eprint/19298/.

Coggan, Andrew. *Unity Game Audio Implementation: A Practical Guide for Beginners*. Abingdon, Oxon: Routledge, 2022.

"Gaming Worth More than Video and Music Combined". *BBC News*. BBC, January 3, 2019. https://www.bbc.co.uk/news/technology-46746593.

Mohn, Christine, Heike Argstatter, and Friedrich-Wilhelm Wilker. "Perception of Six Basic Emotions in Music". *Psychology of Music* 39, no. 4 (October 27, 2010): 503–517. doi:10.1177/0305735610378183.

Schmidt, Brian. "Game Audio Industry Survey 2019". *GameSoundCon*, March 11, 2020. https://www.gamesoundcon.com/post/2019/09/10/game-audio-industry-survey-2019.

Schmidt, Brian. "Game Audio Industry Survey 2021". *Gamesoundcon*, October 2021. https://www.gamesoundcon.com/post/game-audio-industry-survey-2021.

Semel, Paul, and Trent Reznor. "Vintage Interview: Nine Inch Nails' Trent Reznor". *Paulsemel.com*, 2000. https://paulsemel.com/vintage-interview-nine-inch-nails-trent-reznor-2000/.

Chapter 1

Space Invaders (1978)

Mickey mousing, programmable sound generators, and the birth of interactive game music

About the game

One of the most commercially successful and influential games of all times that helped to launch the golden age of the arcades in the end of the 1970s. Players are thrown straight into the action as they are called to defend earth from never-ending waves of space invaders. The player is limited to horizontal movements to dodge incoming laser projectiles and a single button that can shoot back at the alien armada. Unfortunately, resistance is futile and there is no win condition – you can only temporarily hold the invaders back for as long as possible until they eventually overwhelm you.

Fun facts

Designed in Japan in 1978, the launch of the game created such a mania that the surge in demand for coins to play in the arcades led to a national coin shortage and forced the bank of Japan to quadruple the country's coin supply![1] In the next few years the game became a global pop icon generating $13 billion in sales worldwide (adjusted to today's money) and causing such a hype in the US that the Supreme Court considered banning it.[2]

How did the composer get the gig?

The game was designed and produced entirely by Tomohiro Nishikado, a one-person team working for Taito Corporation. Nishikado created all the art, animations, music, sounds, and programming and even engineered the arcade's hardware!

Composition technique 1 – Visual mirroring (Mickey mousing)

The main theme and entire soundtrack of the game consists of a four-note descending chromatic motif: C-B-Bb-A that is infinitely looped until you

DOI: 10.4324/9781003146872-2

meet your inevitable doom (video example 3). Although the theme is extremely simplistic, it works remarkably well with the game for multiple reasons that might not be directly obvious. First, the constant falling chromatic movement of the melody is mirroring the gradual descending motion of the alien invaders. The technique of imitating visual movements through the use of music is one that film composers have utilized extensively in the past, particularly in animation films, and is usually referred to as *Mickey Mousing*. In video example 4 you can quickly notice how closely the music reflects the visual actions of the main characters in a *Looney Tunes* episode of *Roadrunner VS Coyote*: when Coyote falls off a cliff, we usually hear some descending pattern, if he crashes into a wall the music will respond with a sudden accent, if he walks up a staircase then each step will be synced by matching pizzicato notes, etc.

In addition to the visual mirroring, the main theme is entirely built upon a repetition of minor 2nd intervals that progressively get faster in tempo. Can you think of any other famous themes that utilize such a simple yet effective idea? Only three years earlier than *Space Invaders,* John Williams won an Oscar by imitating the foreboding movement of a great white shark in Steven Spielberg's *Jaws*. The motif was based around a constant repetition of two notes, set a minor 2nd apart, that gradually increased in tempo (video example 5, 0:33–0:50). This simple and memorable thematic idea managed to evoke the sensation of something big approaching faster and faster towards the audience. The audience would not actually see the shark visually, but its existence was implied by the music which was perhaps what made it even scarier. The dissonant interval of a minor 2nd (one semitone) which is used chromatically also helped in evoking a sense of discomfort due to its chromatic nature. In *Space Invaders* the moving shark is replaced by aliens that are flying downwards to annihilate earth, but the result is equally effective. This idea has been employed by composers on multiple game soundtracks since to evoke tension and you don't have to search far to find popular variations of it. The exact same approach can be found in the legendary *Sonic the Hedgehog* platformer that helped establish the Sega Genesis console in 1991 as a proper rival to Nintendo. *Sonic*'s drowning motif uses this technique over and over throughout the franchise's history to create anxiety and unease as the player tries to save their avatar from drowning (video example 6).

Composition technique 2 – Competing with SFX

Space Invaders was one of the first games to have continuous music and SFX at the same time. Achieving a good balance between the two can often be challenging due to the unpredictable timing of SFX events that can clash with the music. To solve this issue, Tomohiro separated the two in different pitch registers. Placing the main theme in a very low bass pitch register (similarly to *Jaws*), not only made it more menacing, but also ensured that there is enough space in the frequency spectrum to cut through the SFX that were higher in pitch. To someone new to game audio this area might initially appear to be a problem of the past, but the reality is that even in the higher tech audio environment of today this is a common challenge in any game that has a busy sound environment. As we shall see in later chapters of this book composers have come up with various creative solutions to this problem depending on the style of music and technology available to them. Nonetheless, the arranging approach of using separate pitch registers is an effective solution that provides contrast and limits masking issues.

Composition technique 3 – Adding tempo interactivity

The final and most important point that makes this minimalistic motif so historically significant beyond its commercial influence, is that it is one of the early pioneering examples of interactive game music. The increase in tempo of the four-note motif is directly linked to the movement speed of the Aliens which is in turn determined by the player's kill count. Each level begins with the Aliens moving rather gently but eventually picking up speed as you start eliminating their fleet hence increasing the tension and difficulty of the gameplay. The music reflects this change by gradually increasing in tempo until it reaches an extreme climax of over 650 bpm when a handful of enemies remain. This direct connection between gameplay and musical tempo not only makes the gaming experience more immersive by adapting to the visual action, it also provides direct auditory information about the state and development of the battle. You don't even have to look at the screen to know approximately how many enemies remain and how close to annihilation (or a new high score!) the player is (Figure 1.1).

The interactivity of the music makes the visual mirroring (Mickey Mousing) technique even more effective as it is a direct response to the actions of the player rather than a pre-determined passive experience. Possibly someone playing the game today might perceive such a connection as comedic rather than angst evoking due to the audio quality of the music but nonetheless the underlying mechanics of *Space Invaders* demonstrate how a simple addition of an interactive musical element can make or break the gameplay experience. Just imagine how ineffective this game would be if the four-note music motif remained in a fixed tempo throughout the game, or even worse if the tempo acceleration would happen irrespectively of the development of the battle.

The connection of musical tempo to a gameplay parameter is a useful interactive tool that composers can rely on to reflect changes in tension. A contemporary example of this can be found in the Finnish 2019 action-adventure game *Control*, which alters the tempo of the battle music according to the number of enemies that are alive and in close proximity to the player.

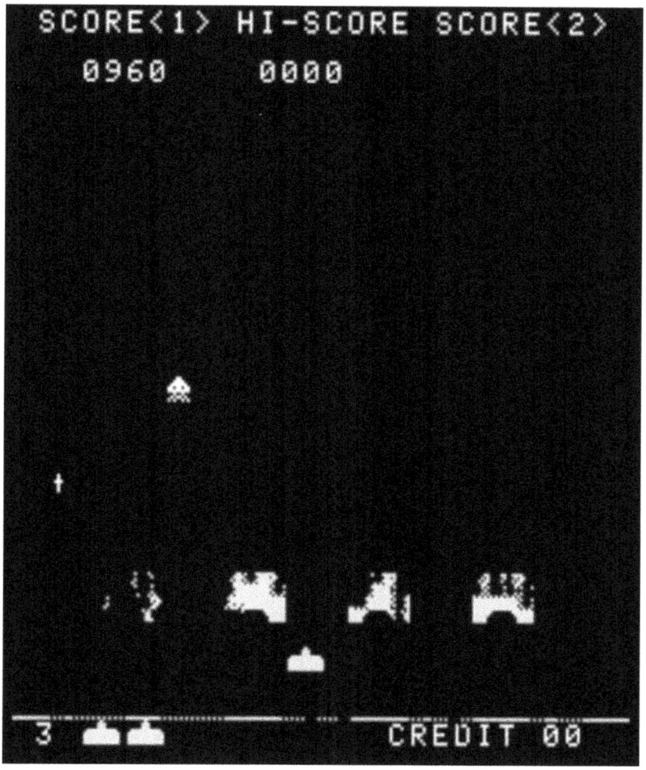

Figure 1.1 A screenshot from *Space Invaders* showcasing a moment in which the music tempo would reach its maximum as there is only a single enemy remaining.

The interesting addition here is that the tempo also works as a tension indicator in the opposite direction and slows down as you start to eliminate most of your foes (see Chapter 19: *Control*).

Production tools – PSGs (programmable sound generators)

From the beginning of the arcade era in the end of the 1970s all the way to the beginning of the 1990s the audio quality of games was severely constrained due to the narrow technical capabilities of the hardware. All the music and SFX were produced by sound chips, also known as PSGs (programmable sound generators) that synthesized audio signals from a combination of basic waveforms, envelopes, and noise. Interestingly, the audio limitations and inaccuracies of those early gaming sound chips defined the sound of that era in a style commonly referred to as chiptune or 8-bit audio (even though it was not necessarily always 8-bit). Each PSG had its own design characteristics (ex: different polyphony, signal flow, effects, etc.) which gave it its own sonic signature, and there is a community of modern-day enthusiasts that replicate their sound through hardware reconstructions and direct software emulators.[3]

The original *Space Invaders* arcade machines used the Texas Instruments SN76477. The chip was used in many other arcade games until it was eventually replaced around the mid-1980s by more capable audio technology. Its main characteristics were:

1) The generated audio was monophonic.
2) It could generate only one square (or pulse) wave or digital noise.
3) It only had an Attack and Decay envelope generator with no sustain or release parameters.
4) Its VCO (Voltage Control Oscillator) responsible for generating the frequency/pitch of the square/pulse wave was unstable and could not produce musical scales accurately as it was mainly designed for SFX generation.[4]

Synthesis 101 – Harmonics

If you are new to synthesis, it will be beneficial to understand how soundwaves work so you can control them more effectively. The waveform of any sound (except sine waves) consists of a combination of many tones that vibrate at different speeds measured in frequencies. What defines the timbre of any instrument from another (ex: a piano from a guitar) is the frequency content of each soundwave, along with how these frequencies behave over time (ex: how long they are sustained). The base tone is known as the **fundamental frequency** (or the first harmonic) and is generally louder than any of the other frequencies. The **harmonics** are integer (whole number) multiples of the fundamental frequency, and any tone that is a fractional multiple of the fundamental is considered as **inharmonic**.[5] Traditional musical instruments tend to produce sounds that have a clear fundamental frequency (the main pitch) as well as several harmonic frequencies and therefore sound more "musical". On the contrary, SFX and other nonmusical sounds usually consist of mainly inharmonic frequencies and therefore sound "noisier".

For example, if you pluck the A string of a guitar it will typically vibrate at a fundamental frequency of 440 Hz per second. However, another part of the string will vibrate at twice the speed (880 Hz) of the fundamental but at a lower volume. This tone is known as the second harmonic as it has a 2:1 ratio to the fundamental harmonic and it is interesting to note that this ratio produces the interval of an octave. At the same time, some noise from your guitar pick might produce a few tones that are inharmonic compared to the fundamental (ex: 453 Hz is not a whole number multiple of 440 Hz) and therefore sound "noisy" but are still important in defining the character of the sound. What sets the guitar tone apart from the same 440 Hz A note played on a piano string, is exactly the number, volume, and behaviour of all the extra harmonic and inharmonic frequencies that are produced on top of the fundamental. When game composers want to create a SFX or a synth sound that imitate a traditional instrument (ex: strings), they will start with one of the basic waveforms and then manipulate their harmonic and inharmonic content over time to shape the sound to their liking.

Takeaway tasks

Task 1 – Composition (moderate) – Mickey mousing

Write a motif that imitates the visual movement of an animated game character such as moving closer, falling, jumping, sneaking, etc. The audio-visual synchronization does not necessarily have to be airtight with the animation; the aim is that the music should evoke the corresponding sensation of movement. How would you know if your motif is effective in mirroring movement? Play it to a friend without any visuals or context and ask them to describe the action they are imagining!

A note of caution: the first inclination is often to compose something that directly mirrors visual movement. However, this can become predictable and obvious quite fast, and it is harder to maintain musical interest. It is therefore usually more effective if the music mirroring is slightly more imaginative and sophisticated in its execution. For example, while imitating a visual falling motion, instead of a straightforward scale run downwards, try using a falling pattern that repeats from a different descending note of the scale. As a source of inspiration, you can watch early Disney cartoons such as *Fantasia* (video example 7) and observe some truly imaginative and creative uses of this technique. The most impressive results can be found in moments in which the musical aesthetic remains engaging and flowing rather than a series of loosely joined motifs. Of course, this is easier said than done and *Fantasia* was animated to pre-existing music, rather than the other way around, but this skill can be useful for every game (and film) composer to develop.

Task 2 – Production (easy) – PSG sound chip emulation

Imitate the sound of an early arcade Texas Instruments SN76477 PSG by using a virtual synth. You can create a new patch in the synth of your choice (ex: Alchemy in Logic, Analog in Ableton, or use Helm, a free plug-in synth for any DAW) and set it up to following rules as close as possible:

- A monophonic square wave oscillator that has unstable tuning (as it was designed for SFX)
- A Sweepable Digital Noise Generator
- A Filter only for the Digital Noise Generator
- A single Envelope Generator that only has Attack/Decay and no Sustain or Release

Task 3 – Career development (very hard) – Make a game clone

This is a very ambitious idea but what if instead of chasing after game developers you would make your own game? Actors and college roommates Matt Damon and Ben Affleck, while struggling to find acting jobs decided to write their own screenplay, *Good Will Hunting*, which led them straight to the forefront of Hollywood's attention and won them an Oscar.[6]

You don't have to create the next blockbuster, even making a simple game clone will provide a great learning experience as it will put you into the shoes of game developers themselves, as well as give you the chance to showcase your composition and implementation skills within an interactive project. You can use a game project template from UE5 or Unity as a starting point and build upon from there. There is a vast amount of online free asset libraries where you can download art, animations, level objects, or whatever else you might need. Depending on your chosen game engine you need to be comfortable with tweaking pre-existing blueprints (UE5) or copy/paste and altering some basic C# scripts (Unity). Extra points if you can be the next Tomohiro Nishikado and eventually manufacture your own arcade hardware!

Notes

1 "Space Invaders."
2 Hansen, *Game On!: Video Game History from Pong and Pac-Man to Mario, Minecraft, and More.*
3 https://www.blackcatsystems.com/software/Synth-76477-SN76477-Complex-Sound-Synthesizer-VCO-LFO-SLF-Modulation-Modulator-Noise-Oscillators-Sine-Triangle-Sawtooth-Pulse-Waveform-Generators-ADSR-Envelope-Control-Custom-Sound-Effects-MIDI-One-Shot.html
4 Henry, "SN-Voice."
5 "Tones, Overtones, Harmonics, and Partials."
6 Goldman, "Interview: Matt Damon."

Bibliography

Goldman, Steven. "Interview: Matt Damon". *The Guardian*, 2007. https://www.theguardian.com/film/2007/aug/10/1.

Hansen, Dustin. *Game On!: Video Game History from Pong and Pac-Man to Mario, Minecraft, and More*. New York: Palgrave USA, 2019.

Henry, Thomas. "SN-Voice". *Birthofasynth.com*. Accessed 30 September 2022. https://www.birthofasynth.com/Thomas_Henry/Pages/SN-Voice_main.html.

"Space Invaders". *Museum of the Game*. Accessed 30 September 2022. https://www.arcade-museum.com/game_detail.php?game_id=9662.

"Tones, Overtones, Harmonics, and Partials". *Apple Support*. Accessed 30 September 2022. https://support.apple.com/guide/logicpro/tones-overtones-harmonics-and-partials-lgsife4183a5/10.7.3/mac/11.0.

Chapter 2

Ballblazer (1985)

Algorithmic guitar solos to infinity!

About the game

Ballblazer is a one-on-one strange futuristic sports game where you try to score a goal by flying a 3D spaceship around a chessboard that has no visible boundaries. The screen is split in two opposing perspectives and the goal posts change in size as the game develops. It can be played as player vs player, player vs computer, and amusingly computer vs computer.

Fun facts

This is the first game ever developed by LucasArts, the game studio founded by George Lucas in 1982. Among the studio's many extraordinary achievements, is that its graphics department eventually gave birth to Pixar animation studios![1]

How did the composer get the gig?

Although some sources such as Wikipedia reference Russell Lieblich as the composer (he was actually the music programmer), the original music was composed by the team leader Peter Langston. Langston, a programmer by trade, was the first person hired to start Lucasfilm's game department and was responsible for setting up the team that developed the company's first two games: *Ballblazer* and *Rescue on Fractalus!*[2]

Composition technique – The riffology algorithm

The game's main theme *Song of the Grid* is a pioneering example of game music as it is one of the first uses of an algorithmic system that generated new musical variations rather than relying on repetitive loops of the same theme. Peter Langston, being a skilled programmer as well as an algorithmic music advocate, developed and published multiple interesting techniques of algorithmic composition during his career including an obscure

DOI: 10.4324/9781003146872-3

music generation system that utilised 100 public telephone lines connected over a network of synthesizers with a cost of $15,000 dollars.[3] The theme of *Ballblazer* makes use of one of his generative techniques termed as the *riffology algorithm*: a system that makes dynamically weighted choices for the generation of various musical parameters based on a model of human improvisation.[4]

The *riffology algorithm* in this game is modelled around a slightly (according to its creator) lazy musician who is playing a never-ending evolving jazz/rock guitar solo. If you are interested in the technical implementation of it you can find information about the entire code written by Langston at the end of this chapter. However, no programming knowledge is required to understand this design in a conceptual level and the purpose of this chapter is to introduce you to the workflow of such a system, a process that can be designed by any composer outside a software environment (see task 2).

To generate this endlessly varying solo, our guitarist begins by choosing one out of the 40 riffs in her repertoire. These are basically eight note motifs based on the A heptatonic (seven tone) blues scale (A B C D D# E F G) with many of them being heavily inspired by famous jazz riffs of the past. After choosing a riff, she is then presented with a simple set of musical improvisation rules to follow. These rules include decisions such as: how fast and how loud to play each riff, when to omit or merge notes, when to pause for a rhythmic break, and other similar but simple musical choices. To make things a little more interesting these decisions are taken with the use of dynamically weighted probabilities, meaning that the probabilities themselves are not fixed but are altered by other processes as the song develops. For example, the probability that controls the energy and tempo of the guitar solos is frequently altered *to imitate a sense of musical development.* At the end of each riff, the guitarist will then pick another one from the database and the process will begin again from the start until it is stopped by the game system.

An interesting point to note here is that the system would also take into consideration how the last and first notes of each riff are related to make sure that transitions between them are relatively smooth, just as a real guitarist will do while soloing. Langston had to make sure not to limit the possible transitions too much (ex: by only connecting riffs which are too similar to each other) as this would make the outcome more deterministic by eliminating a large number of possible combinations. This is a difficult compromise that algorithmic composers often have to deal with as musical coherence often comes at the price of a more predictable output, at least in relatively basic generative systems.

As you would expect with most jazz bands, there is also someone playing the bass, the drums, and the chords which are all also generated and controlled by the same mechanisms. However, the accompaniment uses a simplified version of the *riffology algorithm* based on longer phrases of four bars and produces more reliable but less varied results. It is interesting to observe that although the system has no harmonic awareness at any level, the outcome remains relatively convincing. This is achieved by composing a harmonic progression that would be relatively consonant with any melody that uses the blues scale. Langston does not mention which possible chord sequences were allowed but a safe guess would be a typical I, IV, V structure that is typically used in 12 bar improvisatory blues music.

Take a listen to video example 8 that showcases 33 minutes of generated music from this system.

Listening to the recording we can observe that the composition does achieve the aim that Langston intended: "an infinite, non-repeating improvisation over a non-repeating, but soon familiar, accompaniment".[5] The switches in the probability values as the piece evolves do indeed create a certain sense of development and contrast especially if you compare sections that are a few minutes apart. Langston points out that although the final result sounds musical, it is not always particularly interesting and suggests many areas that the algorithm can be further improved: (1) making the program track harmonic motion in order to allow a greater variability in chord sequences, (2) use riffs of different lengths, (3) have a more complex rhythmic structure, (4) take into account the guitar finger positions, and others.[6]

Production tools – The POKEY PSG

As many other pioneering examples of game music of the time, this technique generative approach was developed as a creative response to the constraints of memory and disk space available in the gaming technology of the time. To understand the difference of scale with today's technology, the ZX spectrum for which this game was ported, only used 16 KB of RAM, while a modern-day affordable gaming PC can easily have 16 GB of ram, a size factor of a million to one!

The Atari systems that *Ballblazer* was designed for used a sound chip with a unique sound quality, called POKEY, that was also utilised in many arcade machines of the time. What gave POKEY its characteristic bright and rich timbre was its inability to maintain an accurate pitch, primarily when all of the four available voices were used simultaneously to play the same note, leading to an unexpected chorus effect. Moreover, its lack of a low pass filter and its ability to produce different types of distortion further added to its appeal for chiptune enthusiasts.[7] It is interesting to note that some arcades used multiple POKEY chips in a single system and *Ballblazer* was one of the few games that actually had a POKEY chip *embedded on its own game cartridge* in order to improve upon the original sound of the Atari 7,800 series.[8]

The game was ported to different platforms that utilised a number of sound chips each producing a different sound. You can listen to a rendering of the theme across all of them in video example 9.[9] Some highlights to focus on are as follows:

1) The ZX Spectrum which only had one channel available making it impossible to play both the guitar solo and the accompaniment at the same time!
2) The Apple II strange synthesis system that relied primarily on clicks
3) The Atari systems POKEY and their rich chorus square waves

Synthesis 101 – Common synth waveforms

The PSGs of the 1970s and 1980s usually relied on one or multiple channels of the following types of soundwaves: sine, triangle, square/pulse, and noise (see Figure 2.1).

Sine

A sine wave is the purest form of sound, it occurs when a sound wave contains only a fundamental frequency with no other tones. Such a sound does not occur in the natural world but can easily be replicated with a synthesizer. According to the Fourier Theorem, all sound can be broken down into individual sine waves. The sound of a sine wave is very soft and quiet, and the lack of other tones makes it easy to integrate in a mix without causing masking problems, making it ideal for low frequency sounds such as bass or kick drums.

Triangle

A triangle is a soft sounding wave, but it is a little harsher and punchier than a sine as it has a series of odd number harmonics (ex: 1, 3, 5, etc.) on top of the fundamental. It is commonly used for bass, chords, as well as soft sounding melodic instruments (ex: synth flute).

Pulse/square

A square wave has the same harmonic structure as the triangle, but the amplitude of each harmonic is much louder, resulting in a harsher, richer sound. This is great for lead melodies as the sound is very dominant. In many examples of game music of the earlier PSG era, composers chose to double the melody using two square waves. This made the sound especially thick and generated a chorus type of effect, caused by the inaccurate synchronization of the frequencies of each oscillator.

Noise

Noise generators consist primarily of many inharmonic overtones which are incredibly complex for our brain to distinguish their harmonic relationship and therefore we hear them as a clatter of sound. This is useful for designing SFX and percussive sounds (ex: snare). There can be different type of noise waves such as white noise that consists of every possible frequency audible to humans (from 20 Hz to 20K Hz). Noise is the only type of synth waveforms that can be easily observed in nature. Next time you are on the beach, just pay attention to the sound of the sea; there are thousands of different frequencies generated every time you hear a splash!

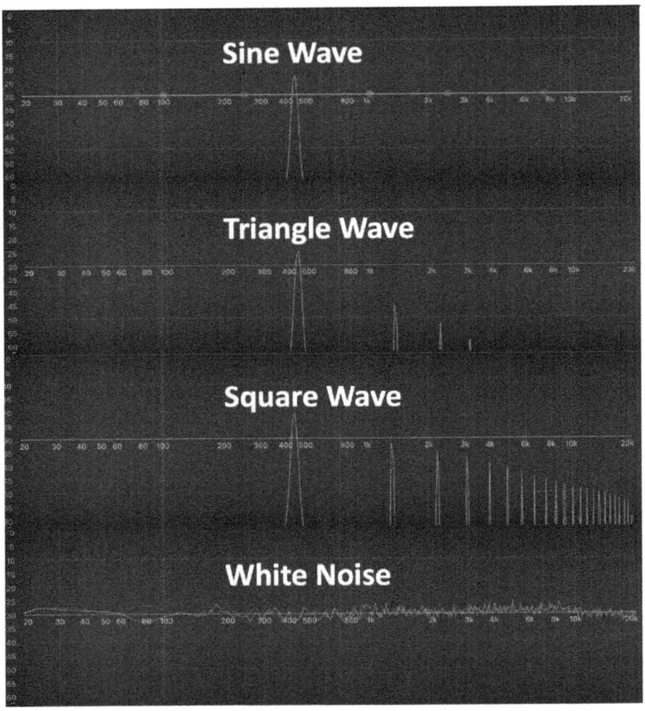

Figure 2.1 An EQ analyser showing the frequency content of the note A (440 Hz) produced by four different synth oscillators. Notice that the sine has only one harmonic, the triangle and square have identical harmonics but in different volumes, and the noise has an extremely dense frequency content with no distinct harmonics.

Takeaway tasks

Task 1 – Research (easy): Identify other algorithmic techniques

Algorithmic composition is not only limited to video games. There is a ton of research available in the field of computer music that has not made its way to games which could be an exciting pathway for the future of video game music. A good starting point if you are interested in this area is to read Langston's paper "Six Techniques for Algorithmic Music Composition" which contains part of the code for the *Ballblazer* algorithm.

Task 2 – Composition (medium): Create an algorithmic flow chart

While having programming knowledge is necessary to be able to practically realize your algorithms, you can still design such systems on a conceptual level. An algorithm is simply a well-defined set of instructions, similar to a precise recipe that is used to accomplish a specific task. To take your first steps in algorithmic composition all you have to do is to create a set of specific rules that will govern the generation of your composition. One of the easiest ways to achieve this is to use what is commonly known as a flow chart: a type of diagram that visually represents an algorithm. You can do this in Microsoft Word or there are multiple free apps that you can find online. There are some simple rules about the use of different symbols but they are not that important at this stage.

In Figure 2.2 you can observe the flowchart I designed for one of the levels in *Apotheon* that is similar to Langston's *riffology* algorithm but even simpler. It plays up to five different motifs according to a fixed probabilities chart but also adds a random delay time before each motif begins which results in a more unpredictable and chaotic arrangement. In addition, it also uses a small dynamic parameter – *Space Invaders* style – that adjusts the volume of the percussion layer based on the distance of the enemies. While such a system has a rather limited output on what type of results it can generate, it still adds an element of unpredictability and variation that in my opinion keeps the music engaging for longer as opposed to a short loop of the same material. You can listen to the result in video example 10 from 0:37 onwards while noticing how the percussion layer changes according to the distance from various enemies.

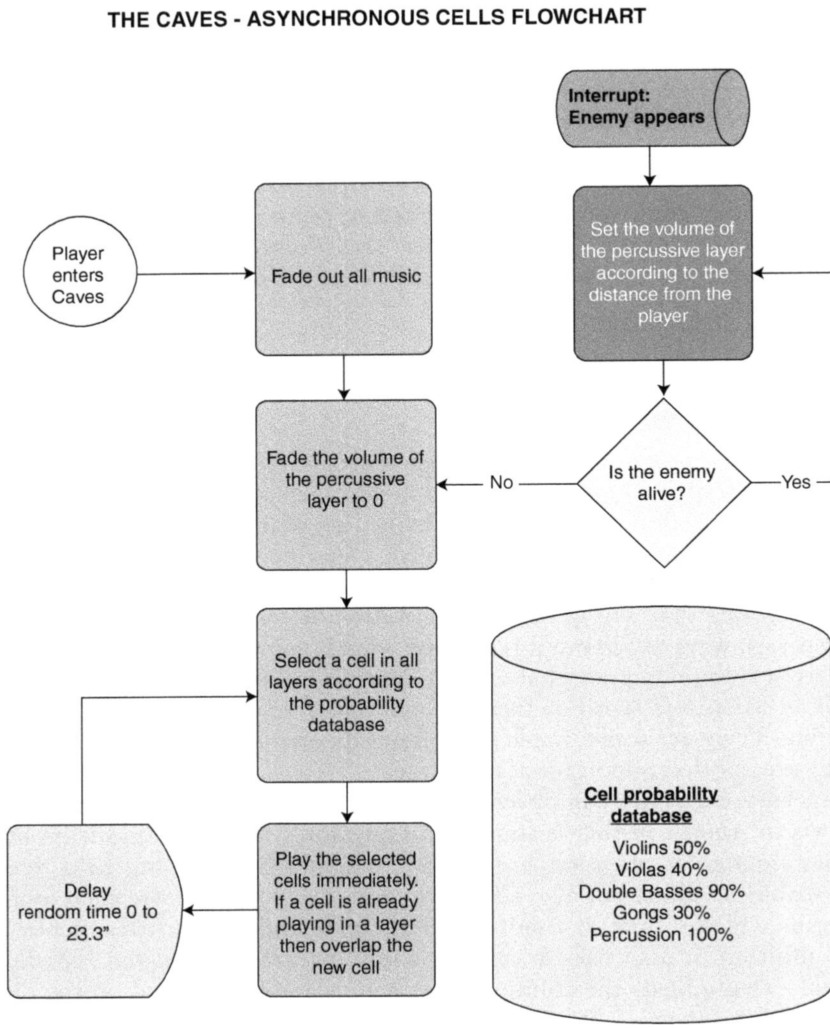

Figure 2.2 An example of a flowchart I designed to help me visualise how the generative music will work for the Caves level in the game *Apotheon*.

Notes

1 "Our Story."
2 Langston, "BALLBLAZER and Rescue on Fractalus!."
3 Langston, "Six Techniques for Algorithmic Music Composition."
4 Langston, "Six Techniques for Algorithmic Music Composition."

5 Langston, "(201) 644–2332- Eedie & Eddie On The Wire, An Experiment In Music Generation."
6 Langston, "(201) 644–2332 –Eedie & Eddie on the Wire, an Experiment in Music Generation."
7 "POKEY."
8 "FAQ Atari 400 800 XL XE: What Are SALLY, ANTIC, CTIA/GTIA/FGTIA, POKEY, and FREDDIE?."

Bibliography

"FAQ Atari 400 800 XL XE: What Are SALLY, ANTIC, CTIA/GTIA/FGTIA, POKEY, and FREDDIE?". *Atarimania.Com.* Accessed 30 September 2022. http://www.atarimania.com/faq-atari-400-800-xl-xe-what-are-sally-antic-ctia-gtia-fgtia-pokey-and-freddie_14.html.

Langston, Peter. "(201) 644–2332 – Eedie & Eddie on the Wire, an Experiment in Music Generation". In *Usenix Association Meeting.* Bell Communications Research, 1986.

Langston, Peter. "BALLBLAZER and Rescue on Fractalus!". *Langston.Com*, 2005. http://www.langston.com/LFGames/.

Langston, Peter. "Six Techniques for Algorithmic Music Composition". In *15th International Computer Music Conference (ICMC)*, 1989. http://www.langston.com/Papers/amc.pdf.

"Our Story". *Pixar Animation Studios.* Accessed 30 September 2022. https://www.pixar.com/our-story-pixar.

"POKEY". *Electronic Music Wiki.* Accessed 30 September 2022. https://electronicmusic.fandom.com/wiki/POKEY.

Chapter 3

The Legend of Zelda (1986)

Music sequences, musical SFX, and the SNES sound

About the game

The Legend of Zelda is a fantasy action-adventure game that follows the adventures of an elf-like boy named *Link*, and *Princess Zelda*. After 19 instalments so far, it is considered as one of Nintendo's most loved and commercially successful franchises.

Fun facts

Composer Koji Kondo wanted to use Ravel's Bolero as the game's main theme as it matched perfectly with the opening screen. However, as he suddenly found out that the copyright of the music had not yet expired (it expires 70 Years after the composer's death) he came up with the legendary Zelda theme in a single night![1] If you compare the two compositions side to side you can find some similarities, notably the rhythm of the accompaniment and the tempo.

How did the composer get the gig?

In 1984, young graduate Koji Kondo searched his school's job placement board to look for his first job. He only applied to work at Nintendo that seemed to be the right fit for him and was hired as the first person in the company to specialize in composition. One year later, he wrote the music for the original *Super Mario Bros* soundtrack, and the year after, the music for the original *Legend of Zelda*. He has remained at Nintendo for his entire life, currently supervising and consulting the *Nintendo Sound Team*.[2]

Composition techniques I – Music sequences

The Legend of Zelda features many memorable melodies that remain widely popular and are frequently performed almost 40 years after its original release. One of the techniques that Kondo frequently utilized in his writing is melodic sequences. This is a simple technique borrowed from early classical music in

DOI: 10.4324/9781003146872-4

which a motif is repeated sequentially but in a higher or lower pitch. If the subsequent repetitions are exact transpositions of the original, a sequence is called real, but if the notes are altered to match the scale, it is called tonal.

The use of melodic sequences can be an easy way of developing and unifying your melody but be aware that overusing this technique can make your musical development more predictable. To avoid this challenge, Koji Kondo often introduces an element of surprise in his sequences by occasionally varying small parts of different elements such as the harmony, the number of repetitions, the melodic direction, and the sequence length.

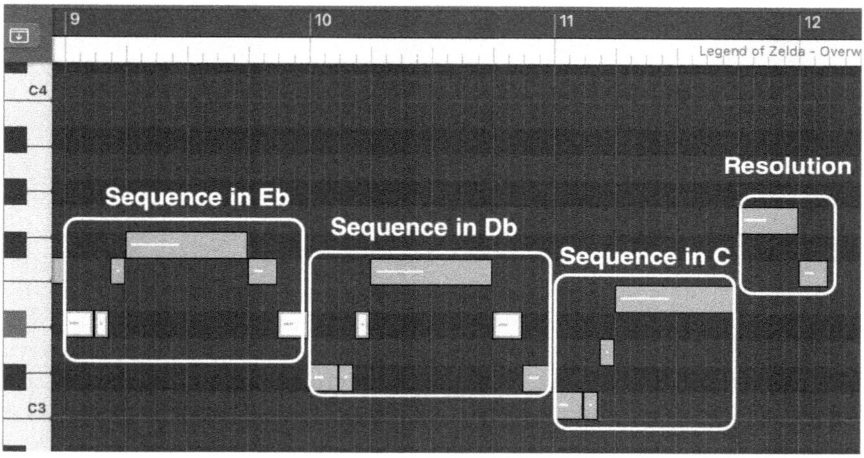

Figure 3.1 An example of a one bar tonal sequence in the Zelda – Underworld theme.

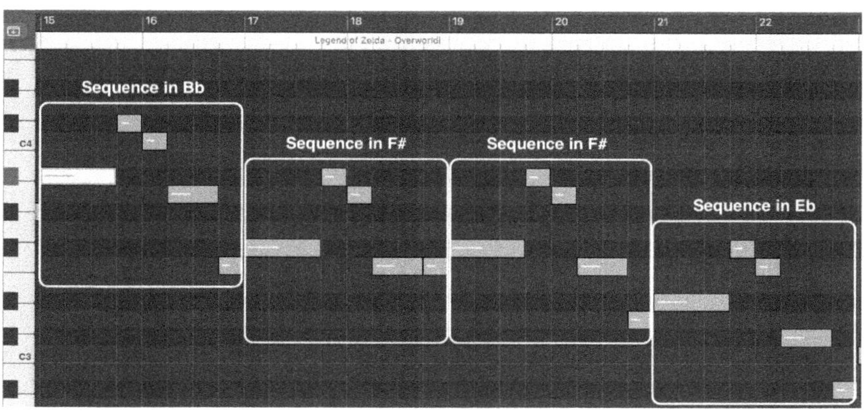

Figure 3.2 An example of a two-bar tonal sequence in the Zelda – Underworld theme.

Listen to the melody of the Overworld theme in in video example 11 while looking at Figure 3.1. The one bar phrase you can hear in 0:13"–0:18" starts at Eb and is repeated two additional times, with each new segment of the sequencing starting from a lower note. Notice that while the rhythm is identical, the interval between the second and third notes of the motif has been increased to two semitones apart rather than one, and the rhythm of the last note has been slightly varied, therefore making it a tonal sequence rather than a real sequence which would be identical. Similarly, you can observe the use of a two-bar sequence in Figure 3.2 from the same theme that repeats from 0:22" to 0:35". The sequence is held twice on the same note before moving on, and again there are some small rhythm and pitch alterations to introduce an element of unpredictability to the development.

Composition technique 2 – Musical SFX

Koji Kondo was responsible for creating both the music and the SFX in the game and allegedly spent an equal amount of time designing both. However, because the sampling capabilities of the original Famicom/NES sound chip were rather limited in reproducing recorded sounds accurately, he chose to give some of the SFX a musical quality. These "musical SFX" accompany some of the main actions in the game and out of a total of 28 SFX in the game, 12 can be perceived as having an identifiable melodic pattern that is generated by the music channels of the sound chip, while the remaining 16 are more conventional SFX that are generated by the sampling and noise channels. The different use of sound chip channels becomes evident when music and multiple SFX are triggered simultaneously which results on audio glitches caused by the inability of the sound chip to reproduce both when it reaches its maximum number of voices.

Listen to all the SFX from the game on video example 12 while observing the score transcriptions in Figure 3.3. As you can quickly tell, a very interesting design approach that is still utilized in contemporary games is that the SFX have noticeably different musical characteristics depending on if the action they accompany is positive (reward) or negative (punishment). For example, one of the most recognizable positive sounds is the *Chest Opening/Item received* sound. The music mirrors Link's movement with an upward chromatic motion as he raises the item above his hands. This ascending motion as a response to a positive action is repeated in all the "reward" type of SFX: the *Rupee get* sound is simply an ascending perfect fifth; the get Heart (life) is an ascending perfect fourth; *open door* is an ascending arpeggio, while defending Ganon, the final boss of the game, produces a rising arpeggio that resolves into a chromatic run. This melodic convention is reversed for any actions or events that are considered negative which are represented by descending motions. The most elaborate example is the *life lost* SFX in which you have the well-known motif of a fast-descending chromatic arpeggio; *enemy hit* is a very fast descending tritone.

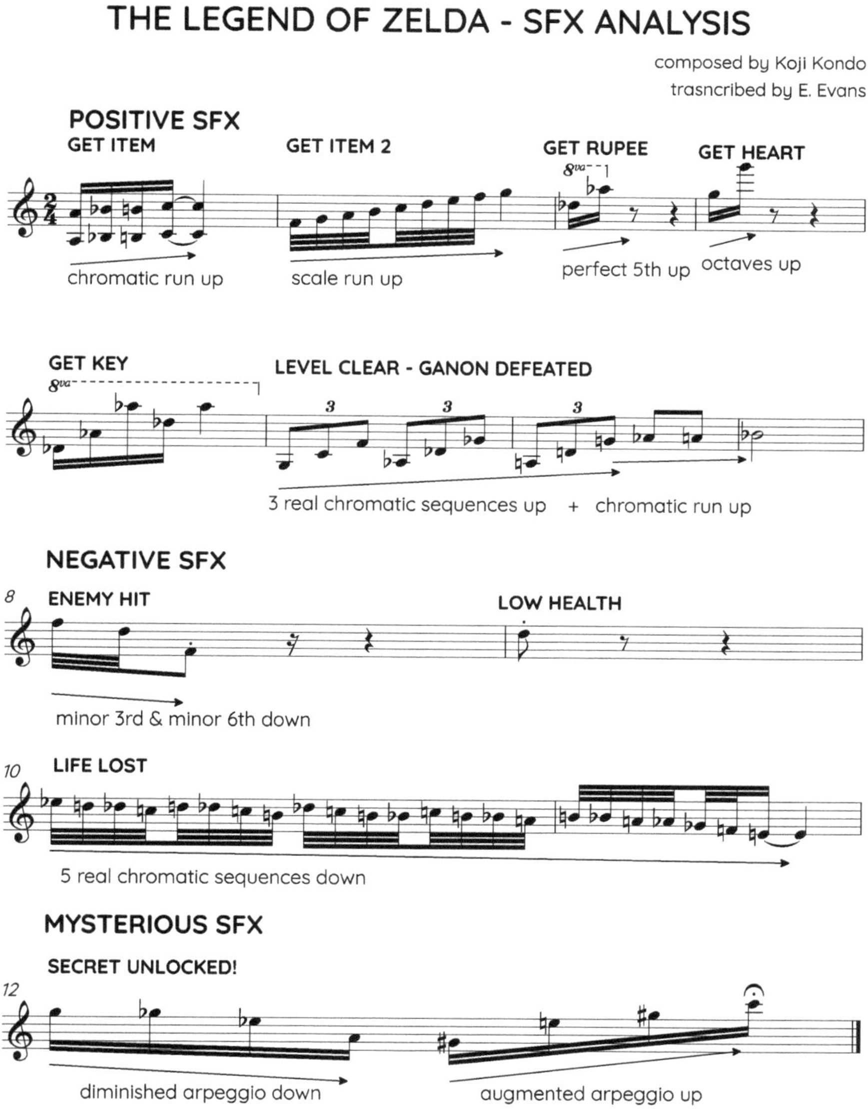

Figure 3.3 Transcriptions of some of the key Musical SFX in *The Legend of Zelda*.

The most famous SFX from the game that will immediately be recognized by anyone who has played any Zelda game is the mysterious *Secret Found*. The association of positive action/ascending motion and negative action/descending motion is combined here with the melody first descending and then ascending, perhaps to convey a more mysterious message to the player as many of these secrets might be challenging to solve. This approach of using "musical" SFX as an element of sound design can be observed throughout Nintendo's subsequent history and some of the original Koji designs have withheld the test of time as they remain present in some form in many of the Zelda sequels to date. For instance, the *Secret Found* motif was played by a harp to match the natural sounding orchestral soundtrack of *Zelda Wind Waker* while in the most recent *Breath of the Wild* it is played by a piano with more elaborate harmonic variations. Video 13 shows you the various transformations that the *Get Item* SFX has gone through the franchise. Musical SFX were an important early development in interactive music composition as they function similarly to the musical stinger technique (see Chapter 18: *Shadow of the Tomb Raider*).

Production tools – The Famicom/NES PSG

The Japanese *Famicom* home console and its western *NES* version (Nintendo Entertainment System) were both among the most iconic 8-bit home console systems of the 1980s that helped establish a new era of home gaming. Their integrated sound chips could theoretically produce five sound channels: two pulse waves, one triangle wave, a noise generator (often used for percussion), and a low-quality digital sampler based on a technique called delta modulation. However, due to the limited available RAM in the stock NES, the digital sampler was almost unusable, therefore limiting the polyphony to only three available voices and noise for both music and SFX, a limitation that required creative arranging decisions by Kondo.[3]

In 1986, a peripheral add-on Disk System was launched which connected to the Famicom and expanded its capabilities. Among other improvements, it allowed games to use cheaper floppy disks rather than game cartridges and expanded the music polyphony from three to four notes by adding an additional audio channel that used a new type of wavetable/sampling synthesis that could generate sounds that were closer to the texture of real instruments. *The Legend of Zelda* was initially released only in Japan in 1986 as an exclusive launch game for this system and Koji Kondo took advantage of this more accurate wave generator to create richer and more impressive SFX that were not possible before, such as the laser style sound of the sword attack SFX and the monster appear SFX.[4] Kondo also used these new sound capabilities to add vibrato at moments that there were no sounds present such as during of the titles, game over, and end credits.

A year later than the Disk System release, the game was brought to the stock NES in the West but as the PSG was more limited the audio had to be reduced. If you are interested, you can observe the unique sonic differences between these two sound chips in video example 14. Also, by observing the gameplay in video 15 from 26:45 you can clearly notice how the NES version was unable to replicate both the SFX and music and frequently breaks down when there are both SFX and music simultaneously due to having one less voice available.

Takeaway tasks

Task 3 can be combined with the other tasks or practised separately.

Task 1 – Composition (easy) – Write a theme that makes use of melodic sequences

You might find it easier to use repetitions of tonal sequences that only use notes of the same key signature but have a different starting note (either upwards or downwards). Remember that after a few sequence repetitions, the melodic direction might start to feel a bit predictable, so make sure to balance it with introducing some element of surprise or novelty to maintain interest.

Task 2 – Composition (easy) – Redesign 5 SFX from Zelda using musical phrases

You can choose any sounds you want out of the series, some ideas include: treasure chest open, get coin (rupee), boss defeated, secret found, life lost, game over. Remember to categorize them as either positive/reward, negative/penalty, or neutral and to make sure that the musical phrase clearly communicates this classification to the player. To test the success of your work, play all your SFX to someone else without giving them any context and ask them to categorize each sound using the same action groups (positive, negative, neutral). Aim to think of a creative way to clearly communicate the action but in a way that is also musically intriguing.

Task 3 – Synthesis (moderate) – Write a theme modelled after the stock NES sound chip

Write a theme modelled after the stock NES/Famicom sound chip. As the earlier generation of game composers have repeatedly shown, sometimes limitation can be inspirational! Many game soundtracks of the 1980s like *Zelda* are heavily melodic, possibly as a response to the limited voice polyphony of the gaming systems of the time. You have only the following voices at your disposal: two square waves for melody/harmony, one triangle for bass, and one white noise channel for your percussion. Remember that the maximum polyphony can only be three notes at the same time (not counting the percussion).

Notes

1 Kondo, "NES Special Interview – Volume 4: The Legend of Zelda."
2 Kondo, "Koji Kondo – 2001 Composer Interview."
3 Kondo, "The History of Nintendo Game Music (1983–2001)."
4 Kondo, "The History of Nintendo Game Music (1983–2001)."

Bibliography

Kondo, Koji. "Koji Kondo – 2001 Composer Interview". *Shmuplations.Com*, 2001. https://shmuplations.com/kojikondo/.

Kondo, Koji. "NES Special Interview – Volume 4: The Legend of Zelda". *Nintendo of Europe Gmbh*, 2016. https://www.nintendo.co.uk/News/2016/November/Nintendo-Classic-Mini-NES-special-interview-Volume-4-The-Legend-of-Zelda-1160048.html.

Kondo, Koji. "The History of Nintendo Game Music (1983–2001)". *Shmuplations.Com*. Accessed 30 September 2022. https://shmuplations.com/nintendogamemusic/.

Chapter 4

Amegas (1987)

The birth of the tracker sequencer

About the game

Amegas is a relatively unknown indie game released exclusively for the Amiga in 1987. It was heavily inspired by *Arkanoid*, the popular block breaker arcade game of 1986.

Fun facts

The game has no musical accompaniment other than the *Amegas* main theme that only plays over its main menu and high scores (video example 16). It possibly marks the first game soundtrack to have been created by a new type of music sequencing software called a tracker.

How did the composer get the gig?

German composer and programmer Karsten Obarski developed the first commercial tracker in 1987 named *The Ultimate Sound Tracker*, which he then used to compose the music for this game. He was a friend of the game developer Guido Bartels, who asked him to write a Commodore 64 style music for the game. After *Amegas*, Obarski moved on to compose music for several Amega games before mysteriously disappearing from the scene.[1]

Composition technique 1 – Tracker sequencing in *Amegas*

When you look at the UI of a tracker sequencer such as the one used in this game (Figure 4.1), it might appear confusing at first, but once you get familiar with the basics you will realize that it offers a straightforward way of producing music that can be even quicker than in a modern DAW. The principal differences are that musical time runs vertically rather than horizontally and that musical notes are triggered by a series of text commands rather than MIDI notes. Songs are created using four independent channels

DOI: 10.4324/9781003146872-5

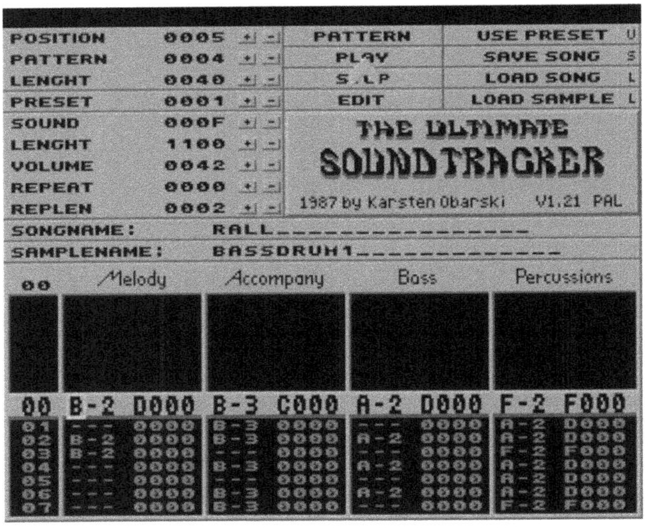

Figure 4.1 A screenshot of the *Amegas* theme in MOD format within *The Ultimate Sound Tracker*, the first commercial tracker sequencer.

(the equivalent of four tracks in a DAW) organized as Melody, Accompany, Bass, and Percussions to match the four-note polyphony of Amega's Paula sound chip.

Let us examine some of the fundamental sequencing techniques that were used in *Amegas*:

Text commands

Looking at Figure 4.2 you can observe how the opening of the *Amegas* theme is sequenced using text commands that trigger specific samples at a specific timing. For example, if you look at the first line of Track 03, step 00, you will notice the following text: "C-206----". The first part of the text indicates the note and octave to be played (C-2), followed by a column that indicates the number of the instrumental sample to be triggered (0), and a final column that indicate any effects to be applied ("----" meaning no effect). Similarly, Track 01-step 0 indicates that a C-2 should be played using instrument 01 (which is happens to be a bass) with no effects applied. When there is no text information in a cell (ex: Track 01, steps 01 and 03) then no new sounds will be triggered during that time unit, thus creating a rhythmic pattern of one note followed by a pause. The interesting part is that the duration of the note is dependent on the length of the sample. For example, in Track 03, step 00, the C-2 note is held as the sample has a long duration but in Track 04, step 00. The sound is staccato as the high hat sample is very short.

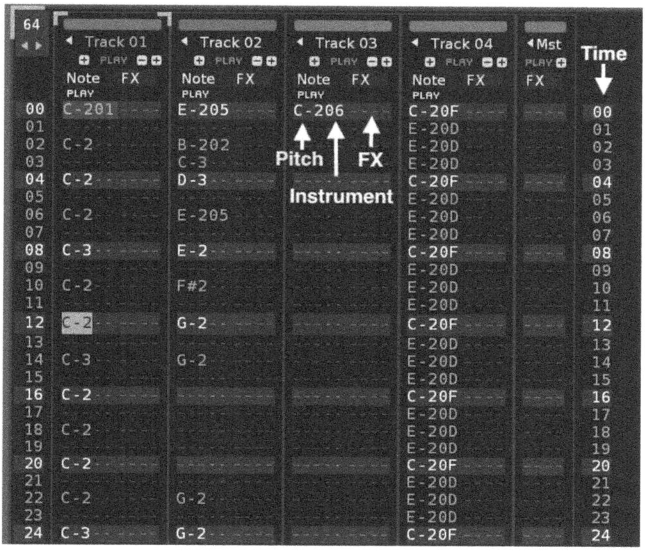

Figure 4.2 A screenshot of the *Amegas* theme in MOD format within ReNoise, a contemporary DAW based on the heritage of classic trackers.

Illusion of polyphony

An important accomplishment with early Trackers is that although you were limited to four channels/voices that did not mean you were limited to only four instruments in your arrangement as you had the ability to switch instruments multiple times by typing the corresponding sample identification number next to the note name. Look at Figure 4.2 again, but focus on Track 02 (which plays the melody) from steps 00 to 06. You might notice that the sample number changes, for example in step 00 the command is to play a E-2 using instrument 05, but in step 02 it is to play a B – 2 using instrument 02 thus creating a countermelody of different instruments within the same track. Switching instruments in the same track through the use of countermelodies can create richer textures as well as an illusion of polyphony without taking up any additional voices as the samples are not triggered simultaneously. This approach was useful in surpassing the sound chip limitation of four voices and was frequently used in this theme. You can watch video example 17 that shows a rendition of the entire *Amegas* theme to observe this technique further, particularly in tracks 02 and 03 that are full of call and response melodies that switch between multiple instruments.

The ability to use multiple instruments in the same channel is one of the characteristic sounds of the Chiptune era that composers have often explored creatively. For example, you can easily change the instrument number next to every other note in a melody creating the illusion of having multiple voices playing, a process that is much more intuitive than in a modern DAW where you would have to spread MIDI notes across an array of instrument tracks at the correct rhythm.

Programming FX

Although no effect commands were used in *Amegas*, subsequent generations of more powerful trackers included many accessible options that could be easily triggered by expanding the text commands with an additional three digits. The exact list of effect commands might differ from tracker to tracker but the general format is usually similar: the first letter digit identifies the effect, and the x/y numerical digits set the amount of the effect. Here are a few examples from ReNoise a modern day tracker that can function as a DAW:

- **Axy** – Set arpeggio, x/y = first/second note offset in semitones.
- **Vxy** – Set vibrato (regular pitch variation), x = speed, y = depth.
- **Bxx** – Play sample backwards (xx = 00) or forwards (xx = 01).[2]

Building a song out of patterns

If you look at the MOD file of the *Amegas* theme (video example 17), you might notice that the song is structured around individual loopable patterns consisting of 64 lines of code each. They are ten unique patterns whose order and number of repetitions is determined by the Sequence List. Using patterns as building blocks is a simple and quick way of building a song as you can easily re-arrange the order of patterns in the sequence list and all note data will be automatically generated to match the new form. Each pattern can also easily be duplicated and edited to quickly create variations. Watching the rendition of the *Amegas* theme you can observe how many patterns are built upon variations of the same code.

Production tools – Contemporary trackers and the MOD format

The Ultimate Sound Tracker played a central role in the development of game music production during the late 1980s and early 1990s. Until that time, game composers would usually write music on a traditional musical instrument and then go through a notoriously difficult programming process before they would be able to reproduce it using the game system's sound chip. However, the unique workflow of the tracker allowed Obrarski to quickly input musical ideas and then immediately test the musical result using the Amega on-board sound chip called Paula. Moreover, as *The Ultimate Sound Tracker* could run as a piece of software within the Amega OS, it allowed for wide access by indie musicians as the cost was very low especially compared to other innovative computer compositions systems of the time such as the Fairlight CMI that would retail for over £25,000! A further advantage was that the MOD file format that Obarski developed to store the music data, was self-sufficient as it included an editable version of the composition as well as the sounds samples required for its performance. Therefore, Obraski had more control over its sonic result especially when compared to other storage formats of the time such as MIDI, in which the outcome would vary considerably based on the General MIDI sounds available in different sound cards (ex: see the different MIDI renditions of *Monkey Island* in Chapter 5).

If you open the *Amegas* Theme MOD file (it is freely available on the internet) you will find a Global Settings file with parameters such as tempo, a set of the ten patterns that contain all the note data, a Sequence list that dictates the order that the patterns should be played, and 11 individual samples each with their own instrument number. Opening or playing the *Amegas* MOD file in any MOD compatible software will result in an identical performance of the composition regardless of your audio gear or sound chip. The original MOD format allowed for up to 15 instruments and four channels of simultaneous playback to match the capabilities of the Amega sound chip, but the format is still available in most tracker software today with expanded capabilities.

Unfortunately, *The Ultimate Sound Tracker* was not a commercial success, but its fundamental design was preserved in many popular Amega and Atari Tracker sequencer spin-offs that followed, as well as many later recreations for Windows aimed at chiptune enthusiasts.[3] The workflow is also strongly present in more powerful tracker software of today such as ReNoise that combine tracker sequencing UI along contemporary DAW features (ex: multisampling, advanced effects, MIDI, VST support). Tracker sequencing offers an alternative producing workflow than a traditional DAW, with its hands-on-keyboard approach that does not rely as much on using the mouse and clicking through menus, along with the vertical use of time and easy pattern creation. This workflow can also be appealing to non-game composers as it can be used to produce a wide range of electronic music using the same design concepts explored in *Amegas* by Obarski. For example, in video example 18 you can see how a more modern EDM track can be fully sequenced and produced in ReNoise.

Takeaway tasks

Task 1 – Remix (easy) – Create your own remix of the Amegas theme

You can use the *Amegas* MOD file as a starting point to create your own remix of the theme in any tracker. There are plenty of free trackers available (ex: Open MPT for Windows) and even the more polished and capable versions like ReNoise offer a free demo. No matter which one you choose, remember that they all function the same way but some individual text commands might differ. The file contains all the original samples and patterns which you can open and edit to create your own arrangement.

Task 2 – Sequencing (challenging) – Create a song using a 4-channel tracker of your choice

For your first track, keep it simple by limiting the composition to just four tracks similarly to the Amega Paula sound chip (Melody, Accompaniment, Bass, Drums). Remember that four tracks do not necessarily equal four instruments like in a traditional DAW and you can interplay instruments at any point.

Notes

1 Borderie, "Soundtracker Origins, Part 1: Where in the World Is Karsten Obarski?."
2 "Effect Commands – Renoise User Manual."
3 For more information on the complex history of the development of trackers and the Amega demo scene you can read McAlpine, Chapter 5, pp. 125–152, *Bits and Pieces: A History of Chiptunes.*

Bibliography

Borderie, Xavier. "Soundtracker Origins, Part 1: Where in the World Is Karsten Obarski?". *Le Weblog De Xavier Borderie*, 2021. https://xavier.borderie.net/blog/2021/09/22/soundtracker-origins-part-1-where-in-the-world-is-karsten-obarski/.

"Effect Commands – Renoise User Manual". *Renoise*. Accessed 6 October 2022. https://tutorials.renoise.com/wiki/Effect_Commands.

McAlpine, Kenneth B. *Bits and Pieces: A History of Chiptunes*. Oxford University Press, 2018.

Chapter 5

The Secret of Monkey Island (1990)

The Secrets of Pirate Reggae!

About the game

The first instalment of the much-loved LucasArts series of point and click adventure games (Figure 5.1). It follows the swashbuckling quests of Guybrush Threepwood, a hopeful but clumsy Caribbean pirate, that must fight the undead, win the affection of the girl of his dreams, and solve the mysterious secrets of Monkey Island!

Fun facts

The "real" secret of Monkey Island has never been revealed although it actually exists. Series creator Ron Gilbert intended to reveal it in the finale of the trilogy but left the company before its completion.[1]

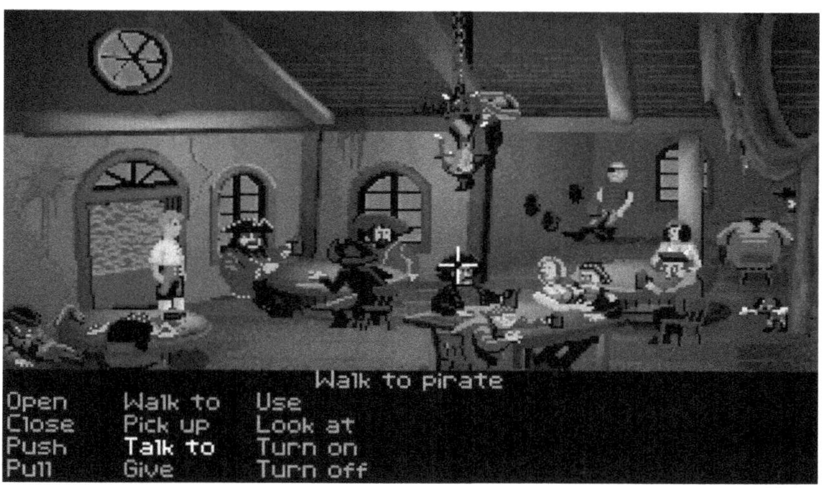

Figure 5.1 A screenshot from *The Secret of Monkey Island* demonstrating the iconic point and click UI that was used in most LucasArts adventure games of the time.

DOI: 10.4324/9781003146872-6

How did the composer get the gig?

Michael Land was writing *MIDI* software as a software engineer for Lexicon Inc. before unexpectedly reading an advert for a composing job at LucasArts in a local newspaper shown to him by his mom. This was the first game he composed music for and despite the rich career that followed (including working in the *Star Wars* and *Indiana Jones* games) it remains one of his favourite scores along with *The Dig*. His frustration with the MIDI system of this game led to the development of iMUSE, an interactive system that synchronizes music with game events that was used in subsequent LucasArts games.

Composition technique – Inventing your own hybrid genre – Pirate Reggae!

The treasured main theme of the series (video example 19) that plays in the opening screen of the game perfectly encapsulates the composer's vision of writing music in his made-up genre of Pirate Reggae, a fusion of classical and Caribbean music. Here are some of the techniques that Michael used that you might find interesting to explore in your own music:

Use of syncopation

One of the most recognizable characteristics of reggae that is strongly evident in this theme is the use of off-beat rhythms and in particular syncopated 8th notes. The easiest way to think of this pattern is to count a regular 4/4 rhythm such as "1 – 2 – 3 – 4" but insert the accented word "and" between each beat, such as "1 – **and** – 2 – **and** – 3 – **and** – 4 – **and**". This rhythm can be expanded further by adding two syncopated 16th notes "ta-ta" at the same place of the "and". Syncopated rhythms such as these are usually played by the guitarist in almost every Reggae song you have ever heard. In Monkey Island, you can clearly hear them in the Marimba and Organ.

For a syncopated groove to really stand out, it needs to be heard in relation to another point of reference that marks the on-beat. In reggae, this is usually achieved by having the bassist play strongly on the beat while the drummer emphasizes the off beats with high hats and snares. The exact same principle is evident in this theme with the bass falling steadily on-beat while the percussion plays in syncopation, mirrored by the xylophone? What is especially interesting is the main melody in the flute which playfully shifts between both on and off beat accents.

Shifting metre

One simple but interesting trick that this song plays on the listener, is that while the music feels familiar and is seemingly easy to follow, the metre has its own secret. If you try to clap along the main theme (video example 19 from 0:21 onwards), you will quickly notice that it is very easy to follow the main pulse, but things might get tricky if you try to count the timing of the melodic phrases. Some bars feel like 3/4, others like 4/4, and some phrases might even fit a 5/4 or a 2+3 count. As the original MIDI file from the game does not contain any information on the metre and there is no official score, there can be multiple interpretations of which division of quarter beats per bar makes the most sense to use for this tune. If you listen to some of the hilarious covers on YouTube (video example 20), you will notice that musicians probably count this slightly differently based on their performances and arrangements. Unofficial transcriptions that are available online also use a range of different time signatures to group the music, with some of them being inaccurate.

In my opinion, the simplest way to interpret the metre without overcomplicating things too much with constant metre changes can be seen on Table 5.1. Melody 1 on the flute can work well enough in 4/4 if you just add two 3/4 bars before and after. Melody 2 is clearly in 3/4 and only the ending of it that functions as a transition is in 4/4, while melody 3 is clearly in 4/4. Lastly, when we return to melody 1 in the end, it still works in a 4/4 metre even though some strong accents on the 4th beat give a 3/4 impression. The use of a shifting metre in this piece establishes an easy to follow but hard to pin down groove that keeps driving the piece forward.

Parallel major/minor

The mood of the piece is clearly happy and humorous but there is one moment in video example 19 (bar 26, 01:00") where the harmony unexpectedly shifts to a darker tone, perhaps hinting the presence of the evil ghost of the notorious pirate captain LeChuck! This is a tonic minor chord (Cm) that is borrowed from the parallel minor that immediately follows the tonic major chord C, thus creating an unsettling feeling as the harmony temporarily switches from the tonic major to the tonic minor by reducing the major third from E a semitone downward to Eb, until it is immediately resolved in the next two bars. It is the same harmonic trick that Gustavo

Table 5.1 The shifting metre of the mysterious Monkey Island theme

Section:	*1 – Free intro*	*2 – Rhythmic intro*	*3 – Melody 1*	*4 – Rhythmic break*
Time:	0:00–0:20"	0:21–0:25"	0:26–0:45"	0:46–0:49"
Bar length:	2 bars	2 bars	8 bars	2 bars
Metre:	Free time	3/4	4/4	3/4
Section:	*4 – Melody 2*	*5 – Transition*	*6 – Melody 3*	*7 – Melody 1*
Time:	0:50-0:55"	0:56-0:58"	0:59-01:09"	01:10-end
Bar length:	3 bars	1 bar	4 bars	10 bars
Metre:	3/4	4/4	4/4	4/4

Santaolalla does in the *Last of Us* theme (see Chapter 11) and can easily be used with switching between the parallel major/minor harmonies.

Free counterpoint

Lastly, another interesting feature of Pirate Reggae is its use of **counterpoint** between different instrumental lines. Counterpoint is a complex area of music theory with a long history dating back to at least the 14th century. There are numerous contrapuntal systems in existence with their own strict rules of contrapuntal motion.[2] The music in the Monkey island theme is using **free counterpoint** which does not follow any specific rules. The key takeaway from this technique is to encourage you to think of your music not only in terms of a homogenous vertical movement but also as a combination of individual horizontal voice movements that have a certain degree of rhythmic and melodic independence from each other, yet are still harmonically co-dependent.[3] As we saw earlier, the theme is filled with musical lines that follow independent rhythms over the same harmony between the bass, percussion, chords, and melody. Moreover, there are also occasional short melodic embellishments that run in between the main melody in different instruments. For example, listen to video example 19 from 0:32" to 0:36" while looking at Figure 5.2, and notice how there are three different melodic lines between the flute, organ, and bass, each following the same overall chord progression but with independent rhythm and voicing.

Production tools – General MIDI

The music of Monkey Island was produced via General Midi and rendered in real time during the game. The sound of each MIDI instrument was dependant on the General Midi sounds that came with the individual soundcard of the system. Therefore, the composer did not have much control of how the music will sound at each home computer other than a choice of instrument names within the General Midi list. You can listen to different renditions of the same theme by a range of sound cards in video example 21, including an impressive performance by a PC with no soundcard at all which plays the MIDI through an onboard sound chip that can only play one sine wave at a time!

Takeaway tasks

Task 1 – Composition (medium) – Write a theme that uses multiple time signatures

A word of caution, if you want the music to be easy to follow make sure to avoid tempo changes. Game and film composers use this technique all the time (ex: listen to *The Exorcist* theme, or the *Silent Hill* soundtrack).

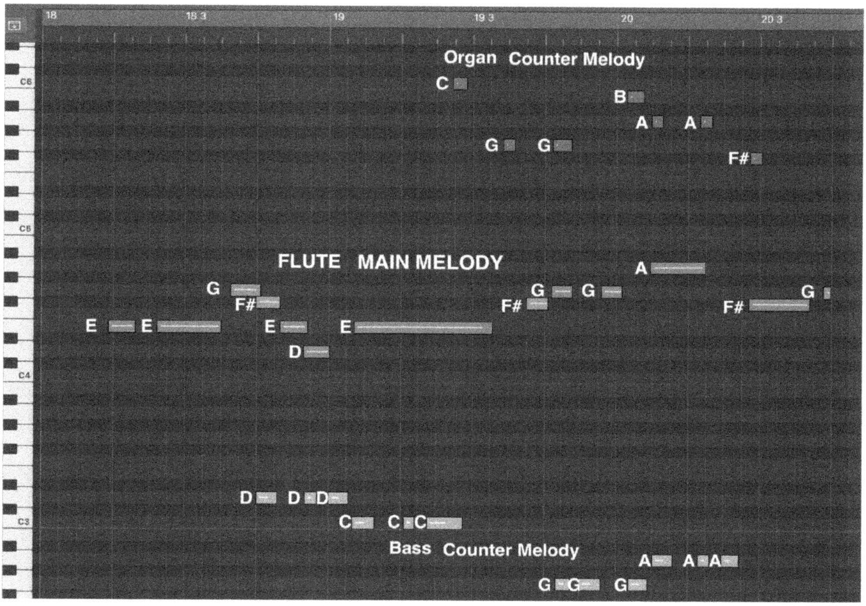

Figure 5.2 2 An example of free counterpoint in *The Secret of Monkey Island* main theme using MIDI transcription. Notice how all lines are harmonically interdependent to each other but they do not move simultaneously.

Take 2 – Composition (difficult) – Write a simple theme that uses a degree of free counterpoint between your musical voices

You can start by setting a simple chord progression over four tracks in your DAW, with each voice playing a different part of the harmony (ex: tonic, third, fifth, octave). Then develop each voice to move independently from each other without altering the harmony by occasionally using different rhythms and melodic patterns. One tip you can try is to include an imitation of one pattern that starts at a later point in another voice. If you enjoy this type of writing, you can have a look at the canons and fugues of J.S. Bach who was one of the virtuosos of contrapuntal writing.

Task 3 – Research (very challenging) – Study the five species of counterpoint

If you are interested in the subject of counterpoint and its use in classical music, I recommend looking into using a cantus firmus (a pre-existing melody) and studying the rules and exercises of the **five species of counterpoint** that were published in 1725 by Joseph Fux and have been used by and directly influenced many renowned classical composers such J.S Bach, Mozart and Beethoven![4]

Notes

1 "25 Fun Facts to Celebrate Monkey Island's 25th Birthday."
2 Sachs and Dahlhaus, "Counterpoint."
3 Laitz, Steven G. (2008). *The Complete Musician* (2nd ed.). New York: Oxford University Press, Inc. p. 96. ISBN 978-0-19-530108-3.
4 Fox and Mann, *The Study of Counterpoint.*

Bibliography

"25 Fun Facts to Celebrate Monkey Island's 25th Birthday". *Grabitmagazine.Com*, 2015. https://www.grabitmagazine.com/blog/post/celebrating-25-years-of-monkey-island-25-fun-facts-you-may-not-know/.

Fox, Johann Joseph, and Alfred Mann. *The Study of Counterpoint.* New York: Norton, 1971.

Sachs, Klaus-Jürgen, and Carl Dahlhaus. "Counterpoint". *Grove Music Online*, 2001. https://doi.org/10.1093/gmo/9781561592630.article.06690.

Chapter 6

Street Fighter II (1991)

Melodic tension in Guile's, Ken's, and Blanka's themes

About the game

Released for arcades in 1991 and home consoles in 1992, *Street Fighter II* was inspirational to numerous competitive fighting games and is still recognizable today. The player engages in combat with eight distinctive characters in street fights taking place around the world before facing the four boss characters.

Fun facts

In the late 1980s and early 1990s the top game composers at Capcom, one of Japan's biggest game development studios, were all women.

How did the composer get the gig?

Yoko Shimomura graduated from college as a classical piano major hoping to become a piano teacher. Capcom was hiring for a composer at the time, so she sent in an application and passed the Capcom entrance exam despite having little composing experience and being discouraged by her music professors. Yoko has had had an amazing career as a game composer with titles such as *Street Fighter II,* the *Kingdom Hearts* series, *Super Mario RPG*, and *Final Fantasy XV.*

Music theory 101 – Nonharmonic notes

Have you ever wondered why the I, IV, and V chords are so frequently used in chord sequences? The combination of just these three chords contains all the notes of a major scale, and therefore, can be used to fully harmonize any melody that makes use of diatonic notes. However, composers might choose to use nonharmonic notes in the

DOI: 10.4324/9781003146872-7

melody that do not match with the underlying chords. Some of the most common uses are:

passing notes – nonharmonic notes that move between two chord notes in a single direction.
auxiliary (or neighbouring notes) – nonharmonic notes that move above or below a chord note but then return to the original chord note.
suspended notes – nonharmonic notes that are held from the previous chord and will resolve by step usually downwards into a harmonic note of the new chord.

Composition technique – Creating melodic tension with nonharmonic notes

Many of the memorable melodies in *Street Fighter II* incorporate diatonic notes that are not part of the accompanying chords. This technique creates an expectation that these nonharmonic notes will be resolved at a later moment and thus drives the music forward with momentum and tension, an approach clearly suited to keeping the player on the edge during the constant fierce street fighting in the game. Let us examine how melodic tension is created in three popular character themes of the game.

Melodic tension in GUILE'S THEME

I personally find this theme as one of the biggest ear worms in the history of game music. I challenge you to listen to an Acapella cover of Guile's theme in video example 21 while studying the analysis in Figure 6.1 and then try to get it out of your mind!

Intro

The song is in the key of Cm (C, D, Eb, F, G, Ab, Bb, C) but the melody does not use the note C not even once during this intro. Instead, it floats around multiple repetitions of Eb and D, until F and Bb are also introduced in bar 3. The note Eb belongs to all three underlying chords (Cm, Abmaj7, Fm) but the note D does not belong to any – it is instead used as an auxiliary note that implies it will resolve to C (the tonic) but keeps returning to Eb. The final two notes (D, Bb) of this phrase are finally matching the chord notes of Gm but the clearly implied resolution of the tension to C does not arrive until the Verse, as the phrase repeats again to prolong the melodic tension for another three bars.

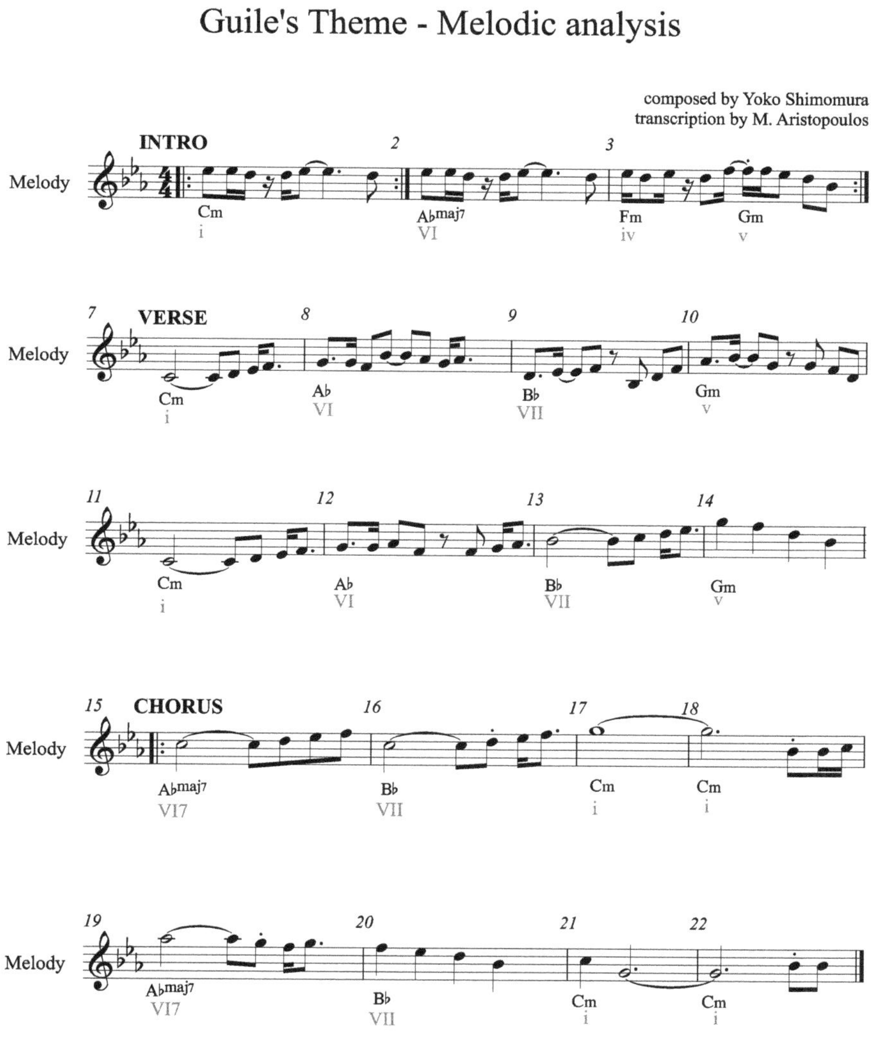

Figure 6.1 A melodic analysis of Guile's Theme. Notice the extended use of non-chord notes in the melody that function as passing, auxiliary, and suspended notes and create melodic tension.

Verse

The melodic tension is finally resolved to C (tonic) in the first note of the verse which clearly establishes the Cm tonality with the use of passing notes between all the chord notes and dismisses any hints towards a move to the Eb relative major. What is noteworthy here is that the verse alternates between a playful game of tension and release from one bar to the next: one bar of harmonic melodic movement of the triad chords is followed by one bar of primarily auxiliary and suspended notes. The melody also keeps rising upwards which adds further to the build-up of tension. However, unlike the intro, the Bb at the very end of the phrase is resolved to C but not in the octave you would expect as it makes an unexpected jump to an octave below to temporarily allow space for building upwards again.

Chorus

The chorus begins with the same melodic phrase of a Cm scale moving upwards as the verse, but it is unexpectedly repeated twice with two new reharmonizations of major chords, which imply that the music might have modulated. However, the key of Cm is quickly re-established as it is repeated for two bars before jumping even higher in pitch to the climax of the top Ab and then finally moving towards a familiar harmonic motion that resolves downwards into C using entirely chord and passing notes.

Melodic tension in KEN'S THEME

This is the theme of the other American character in the game, and it takes the idea of a rock anthem to the maximum. The melodic techniques utilized in video example 23 are even simpler and were extremely popular in rock/metal anthems of the 1980s (just listen to almost any song by Iron Maiden and you are guaranteed to find it). The trick is the following: you take a short melodic phrase and repeat it over a sequence of diatonic chords underneath that harmonically clash with the melody. The dissonance is eventually resolved at the end of the sequence before repeating the entire trick again while usually building up the arrangement along the way. This can work with almost any chords if the first and last chord is consonant, and if the chords are diatonic (they belong to the same key as the melody). You can also potentially use some chromatic chords (see Chapter 8 on *Diablo* for more info) but this might need further planning of the harmonic direction. The easiest approach is to use the bread and butter of rock harmony, the **power chord.** For the uninitiated, a power chord is simply a chord that does not have a third which is the interval that defines its minor or major colour. A power chord is constructed by using only a root tone, a perfect fifth, and occasionally also an octave. Power chords are much easier to implement when building your sequence as there are fewer harmonic combinations that need to be considered.

Melodic tension in BLANKA'S THEME

Yoko Shimomura was struggling to come up with a melody for Blanka's character until she got a strange epiphany while seeing a strange bag with Blanka's colours during her morning train commute to work. The melodic motif that came to her mind did not seem to initially fit with the rhythm section she had previously written as they both used a different key.

The melodic tension and dissonance created by having two different keys playing simultaneously over each part (rhythm/melody) is exactly what makes Blanka's theme so fitting to his unusual character (he is a beast mutant living in the Brazilian jungles), and it is never really resolved in the song (video example 24). According to Yoko:

> Blanka's theme has some really unusual parts. So, when it's arranged, people often end up correcting those parts. The rhythm for Blanka's theme itself is in a major key, but the melody is in a minor key. Basically, you hear an A natural and an A flat at the same time. It's really something that should be fixed, but if I fixed it, it'd become a different song entirely. That strange, broken feeling is what made the song for me. People said the music was wrong at the time, but if so many people tell me they love it now, then I don't think it's wrong. I'm finally able to believe that now.[1]

The idea of using multiple keys at the same time is known as **polytonality** while the use of two keys at the same time as **bitonality**. There are some compositions that briefly used these techniques in the classical music era, but they were really popularized with Stravinsky's pivotal and highly controversial work in *The Rite of String* (video example 25).

Production tools – The YM2151 frequency modulation chip

The original *Street Fighter II* arcade machine had a **YM2151 FM synthesis chip** and a **MSM6295 ADPCM chip** (four channels, used voice synthesis to mimic human speech). It is very interesting to note that the YM2151 was the first single-chip FM synthesizer, and it was made by Yamaha. The chip was originally created for the early Yamaha DX series of keyboards, the precursors to the legendary Yamaha DX7 – one of the most commercially successful professional-level hardware synthesizer of all times![2] The YM2151 chip had 8 voices, 4 operators, and 8 algorithms, compared to the DX7 that had 16 voices, 6 operators and 32 algorithms. Nonetheless, the YM2151 could produce a great number of sounds that were not possible to synthesize with non-FM chips. It did so by only using sine waves with no filters or effects, and was eventually used in many arcade and console game system boards.

Shimomura recalls the complexities of programming music using a sound chip:

> Back when I was composing the SF2 music, I had to make it on a machine with a circuit sound system. We were using a type of FM sound chip, which I think was called YM2151. With that chip we could play the music and adjust it with a program, kind of like an app nowadays. Maybe app isn't quite right, but we had a PC that could run that software basically. At that point we had about a system each for composing songs, so I was composing on my own PC. So, while thinking about

what it'd sound like with the FM sound, I brought in the data, the actual MIDI data, and played it with the FM sound, and then adjusted accordingly. When it didn't play the kind of sound I expected, I'd fix it right there, and since it was an FM sound chip, I could create sounds. Altogether I could save up to 128 or maybe 255 sounds, so I could keep saving them and editing them, then make new ones and edit them again. Ethnic sounds, unique sounds, or something like a guitar is really difficult with FM sound, so I kind of approximated them. And we used a system called ADPCM for the drums only, so we sampled them and played them back, and finally played everything back together.[3]

FM synthesis 101

Frequency Modulation synthesis can be a confusing subject especially with the highly complex capabilities offered by modern FM synths. However, the premise of early FM synths is relatively simple at its core: sounds are produced by having one sine wave (the modulator) modulate the pitch of another sine wave (the carrier). Each sine wave is produced by an oscillator which together with its envelope is known as an "operator". Each FM synth would have multiple operators (ex: the YM2151 had four) that could be arranged in a different order known as an "algorithm".

The primary properties that define the FM outcome are:

1) The frequency ratio between both operators controls the frequency content of the new sound. For example, if the frequencies of the two operators are inharmonic (not direct multiples of each other) the resulting sound will also be inharmonic. Using inharmonic ratios is ideal for quickly producing complex sounds quickly (ex: bells, metallic sounds, synthetic brass).[4]
2) The amplitude (volume) of the first operator (the modulator) controls the amount of FM synthesis to be applied to the other operator (the carrier). For example, the louder you make the modulator the more Frequency Modulation you will get.
3) Each operator is made of a sine wave oscillator with an envelope. My controlling the ADSR of the envelope you can design its amplitude over time (ex: staccato, legato, fade-in, etc.).
4) The order in which the oscillators are connected with each other (known as the algorithm) controls all of the above as it determines which oscillator is acting as the modulator and which as the carrier. For example, the YM2151 had eight algorithms, meaning that its four operators (a sine oscillator with an envelope) could be connected in eight different ways that would produce eight different types of Frequency Modulation.

Takeaway tasks

Task 1 – Remix (challenging) – Create a SF2 remix using FM synthesis

Download the MIDI file from any of the songs in the game (they can be found by a quick google search) but replace the instruments with your own sounds made with an FM synth. There are plenty of powerful FM synths that come with most DAWs such as Operator in Ableton, or EFM1 and Retro Synth in Logic. You can start with just two sine wave operators (one being the modulator, and one being the carrier), and no effects. Then adjust the parameters discussed earlier: the amplitude and pitch of the operator and the ADSR of the envelope. Make sure to experiment with both harmonic and inharmonic ratios between your operators (you can click fixed and set a specific frequency number in Hz in Ableton). If you are using Ableton's Operator, you can then add more operators and make sure to explore the 11 different algorithms. You can also open different pre-sets (ex: Brass) and reverse engineer them to understand how they work.

Task 2 – Composition (medium) – Write a theme that uses melodic tension for one of the original SF2 characters

You can use one or more of the melodic tension techniques discussed in this chapter. Look at Figure 6.2 and the description of each character below to get inspiration:

- **Ryu**: the winner of the previous tournament. A Japanese fighter that seeks no fame but only to develop his Karate skills.
- **E. Honda**: a Japanese sumo wrestler that is strong but slow moving.
- **Blanka**: a Brazilian beast mutant who seeks to uncover his forgotten past.
- **Guile**: a former USA special forces soldier that seeks revenge for the death of his best friend.
- **Ken**: Ruy's best friend but also biggest rival, from the USA.
- **Chun-Li**: a Chinese martial artist seeking revenge for the death of her father.
- **Zangief**: a Soviet Union wrestler seeking to defeat his American opponents with his bare hands.
- **Dhalsim**: a pacifist yoga master from India who seeks to gain money to help the less fortunate.[5]

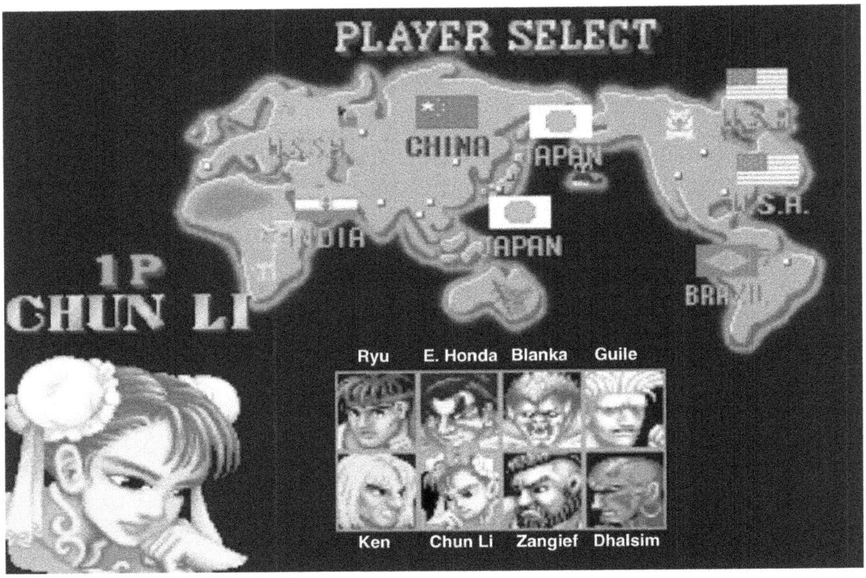

Figure 6.2 A screenshot from the Player Select menu in the original *Street Fighter II* displaying all the available fighter options and their country of origin.

Notes

1 Shimomura and Dwyer, "Interview: Street Fighter II'S Yoko Shimomura."
2 "Frequency Modulation (FM) Synthesis."
3 Shimomura and Dwyer, "Interview: Street Fighter II'S Yoko Shimomura."
4 "Frequency Modulation (FM) Synthesis."
5 "Street Fighter II Characters."

Bibliography

"Frequency Modulation (FM) Synthesis". *Apple Support*. Accessed 30 September 2022. https://support.apple.com/en-gb/guide/logicpro/lgsife418213/mac#:~:-text=FM%20synthesis%20uses%20a%20modulator,range%2C%20thus%20producing%20new%20harmonics.

Shimomura, Yoko, and Nick Dwyer. "Interview: Street Fighter II'S Yoko Shimomura". *Https://Daily.Redbullmusicacademy.Com/*, 2014. https://daily.redbullmusicacademy.com/2014/09/yoko-shimomura-interview.

"Street Fighter II Characters". *Street Fighter Wiki*. Accessed 7 October 2022. https://streetfighter.fandom.com/wiki/Category:Street_Fighter_II_Characters.

Chapter 7

Mortal Kombat (1992)

From the arcades to the dance floor, formulaic writing makes a classic hit

About the game

An all-time classic fighting game released for the arcades in 1992 that shocked the public with its unprecedented levels of fantasy violence and gore. Several fictional themed warriors enter a martial arts tournament and fight to the death for the freedom of their magical realms.

Fun facts

The design was initially inspired by the fighting films of Jean-Claude Van Damme who is imitated in-game by the character of Johnny Cage (Figure 7.1).[1]

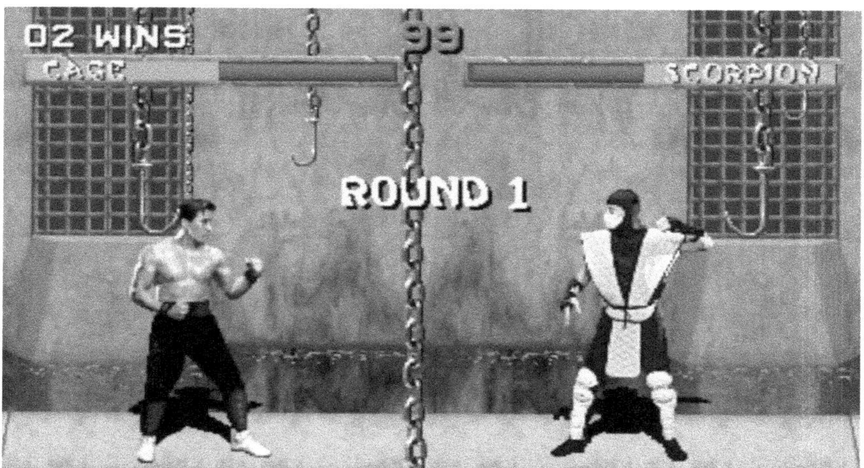

Figure 7.1 A screenshot from *Mortal Kombat* showing the character Johnny Cage that was unofficially modelled after Van Damme.

DOI: 10.4324/9781003146872-8

How did the composer get the gig?

The music of the original game was written by Dan Forden who also worked on the sound effects and was part of the company's design team. However, the famous *Mortal Kombat* franchise theme, also known as "Techno Syndrome", was not part of the first release of the game in the arcades. It was composed by Olivier Adams, a member of the Belgian electronic duo *The Immortals*, when the band was asked to write a TV promotional song to support the game's release into home consoles.[2]

Composition technique 1 – Formulaic pop writing

The composition techniques behind the MK theme song (video example 26) are incredibly simple and formulaic, yet very effective in creating a memorable track inspired by 1990s dance music that managed to become a global hit that unexpectedly reached fans beyond the gaming community and was frequently played on commercial radio stations and dance clubs worldwide. It currently has over 150 million views on YouTube which eccentrically places it at top spot of the most popular game theme of all times on the platform. It has been covered extensively in the last 30 years in a very wide range of genres, from orchestral to heavy metal styles (video example 27), as well as some bizarre versions including medieval acapella groups and even cat singing remixes!

Pop song structure

The structure of the song is based on the well-tested pop song writing formula of a verse-chorus-bridge, along with constant small variations in the arrangement and instrumentation. There are four main phrases, all of them four bars long and in 4/4 that play in the following form: Verse x2, Chorus A x3, Chorus B x3, Bridge x2.

Repetitive lyrics

The lyrics also follow the looping structure of many dance music hits of the time with continuous repetition of a few catchy words and key phrases that become memorable. For example: "Test your might, Test your might, Test your might, Test your might, (scream) MORTAL KOMBAT!"

Four on the floor kick

The drums are built around the bread-and-butter of 1990s dance music, the characteristic four to the floor kick. In contrast to pop/rock where the kick drum hits at beats 1 and 3, while the snare hits at 2 and 4, four to the floor dance music has the kick hitting in all four beats, even under the snare hits.

Following a four-chord formula

Some of the game soundtracks discussed in this book follow a complex harmonic development but this theme is not one of them. It predominately relies on a repeating four-chord progression formula of i – III – VII – VI based in Aminor (Am – C – G – F) that is found in countless pop songs, while the Chorus is only a repetition of an A minor chord, and the Chorus B is mainly based on repetitions of A minor – G – C chords.

If you are not familiar with four-chord formulas you might be shocked when you discover how many popular songs are built upon the exact same chord progression yet sound quite different from each other. In video example 28 (this is a must see!) the brilliant music group *Axis of Awesome* comically demonstrate this concept on a similar four-chord progression of I–V–vi–IV.

Composition technique 2 – Phrygian mode

Most of the MK theme song is based on the key of D minor aeolian which is simply another name of the ascending natural D minor scale: D, E, F, G, A, Bb, C, D. Notice that the 7th is not raised as in the harmonic minor that would have a C#. This is just the same key signature as its relative major, F, but starting on a D, in the same way as A minor is relative to C major.

What gives this theme a little bit of a unique colour that helps set it apart is the descending pattern in the melody of the chorus unexpectedly happens in the mode of D Phrygian video 26, 0:40. The D Phrygian (D, **Eb**, F, G, A, Bb, C, D) is a minor mode that is parallel to D Aeolian and has the same key signature with the only key difference that it also has a flatten 2nd which gives the mode its unique colour.

Parallel modes 101

Using parallel modes is a great way of introducing a different sound to your music. Many music theory books overcomplicate this subject by teaching you how to build modes starting from a different note of the C major scale. While this is useful in quickly remembering the structure of each mode, it is not very practical as all of these will belong to a different key (depending on the starting tone). Modes are much easier to use in parallel, meaning that they should all be transposed to start on the same note. You can easily construct all seven of them by taking any major key and then applying the following formulas:

Table 7.1 The formula for constructing all seven parallel modes starting from a major key and altering specific notes of the scale as needed

Ionian (major mode)	*1*	*2*	*3*	*4*	*5*	*6*	*7*
Dorian (minor mode)	1	2	b3	4	5	6	b7
Phrygian (minor mode)	1	b2	b3	4	5	b6	b7
Lydian (major mode)	1	2	3	#4	5	6	7
Mixolydian (major mode)	1	2	3	4	5	6	b7
Aeolian (minor mode)	1	2	b3	4	5	b6	b7
Locrian (diminished mode)	1	b2	b3	4	b5	b6	b7

Notice that the Dorian, Phrygian, and Aeolian are all minor modes, while the Ionian, Lydian, and Mixolydian are all major modes. The Locrian is a special mode as it resembles the diminished scale and is quite eccentric to use. It is also easier to move from a minor mode to another minor mode (and vice versa for major modes) as the key signatures are more closely related. For example, as we saw in the *Mortal Kombat* theme, moving from D Aeolian to D Dorian only needs one note to be altered, but moving from D Aeolian to D Lydian will need four hence making it a more abrupt change.

Production tools – SFX sampling

Another technique that makes the MK theme so successful is its catchy use of sampling. It might not be directly obvious at first but if you pay attention, you will quickly notice that many musical elements of the production are actually sound effects samples taken from the game itself. There are gongs, punches, kicks, special attacks, character voices, and of course the characteristic *Mortal Kombat* scream that is taken from the Sega Genesis TV advertisement and pitched down an octave. These samples are used musically to create rhythmic patterns, and a lyrical accompaniment, over the rest of the music, and the entire song was produced using an old Atari ST 1040 home computer.[3] This Atari was a fantastic music making machine at the time as it had built-in MIDI ports, it could take floppy disks, and had a large amount of memory that gave it a comparatively strong capability for audio sampling. It is interesting to mention that some of the most popular DAWs of today, such as Cubase and Logic Pro, originated on the Atari ST![4]

Takeaway tasks

These three tasks can be combined or attempted separately.

Task 1 – Composition (easy/medium) – Formulaic writing

Write a theme song that is based on formulaic popular writing techniques. Select a four-chord progression (ex: I-V-vi-IV in C major would be C, G, Am, F) and develop your arrangement around building blocks such as an intro, verse, bridge, and chorus. There is nothing wrong in using such formulas as a starting point for your music but be cautious that if the generation of all your elements is formulaic then the final result most probably will sound generic. The key to making this interesting is to also have something slightly unexpected.

Task 2 – Composition (challenging) write a short theme using a mode

You can easily transition between parallel modes within the same composition by following Table 7.1. An important point to remember is that after constructing a mode to use in your melody, you need to also alter your diatonic chords according by using the same note alterations. For example, if you are using C Lydian that has a #4, you need to alter any chords that have an F to an F#, thus Dm (DFA) becomes a D (DF#A), F becomes a F#dim, and Bdim becomes a Bm. Remember that modes have a very idiosyncratic colour it is rare to use more than two in the same composition and is also much easier to transition between modes of the same quality (major/minor).

Task 3 – Production (easy/medium) – Create an instrument from in-game SFX

You can use the *Mortal Kombat* SFX library (video example 29) or sample any other game of your choice. The *Mortal Kombat* approach of using samples might have a slightly comedic effect with a strong 1990s ambiance but you can also use this technique creatively in other ways. You can use the SFX either through a sampler instrument (ex: Logic – Quicksampler) that is performed in a MIDI keyboard like any other instrument, or you can create instrumental patterns by placing the audio files directly into your DAW session by using a quantized grid.

Notes

1 Kantor and Iannone, "The Untold Truth of Jean-Claude Van Damme."
2 Grebey, Adams and Engelen, "The Team Behind the Mortal Kombat Theme Song Had No Idea They'd Created a Knockout."
3 Grebey, Adams and Engelen, "The Team Behind the Mortal Kombat Theme Song Had No Idea They'd Created a Knockout."
4 Needs, "Dirty Dozen – Micro Music – Jun/Jul 1989."

Bibliography

Grebey, James, Olivier Adams, and Maurice Engelen. "The Team Behind the Mortal Kombat Theme Song Had No Idea They'd Created a Knockout". *Vulture.Com*, 2021. https://www.vulture.com/2021/04/the-mortal-kombat-theme-song-creators-on-their-knockout-hit.html

Kantor, Jonathan H., and Jason Iannone. "The Untold Truth of Jean-Claude Van Damme". *Looper.Com*, 2021. https://www.looper.com/35987/untold-truth-jean-claude-van-damme/

Needs, Paul. "Dirty Dozen – Micro Music – Jun/Jul 1989". *Muzines.Co.Uk*. Accessed 1 October 2022. http://www.muzines.co.uk/articles/dirty-dozen/5214

Chapter 8

Diablo (1996)

Chromatic chords and non-functional harmony in *Tristram Village*

About the game

Diablo is a straightforward hack and slash action RPG that was praised for its highly immersive gothic atmosphere and addictive gameplay design. Players must fight their way through a haunted cathedral filled with demonic forces that descends all the way to hell. The game world is based on a procedural system that gives the game a very high replayability value by generating unpredictable dungeon designs, randomized item drops, and variable quest lines.

Fun trivia

Diablo is one of these rare games that were so influential that it managed to create an entire gaming sub-genre referred to as *Diablo Clones*. The franchise has such a passionate and overdemanding fan base, that after the studio announced that the long-awaited sequel – *Diablo Immortal* – will be released as a simpler mobile phone game rather than as a fully fledged PC/console title, the developers were booed off stage in their own game conference, the company's stock went tumbling, and the promo trailer became one of the most disliked videos in the history of YouTube.

How did the composer get involved?

Matt Uelmen got the gig by persistently cold calling several developers of a small game studio named *Condor Games* that he found on an old *Nintendo* document and offering to create a demo for their games. After working on a few titles together, the studio team started developing *Diablo* before being purchased and renamed as *Blizzard North*. Despite the widespread popularity of the game, the original *Diablo* soundtrack was released 15 years later as an anniversary release.[1]

DOI: 10.4324/9781003146872-9

Music theory 101: Diatonic and chromatic chords

If a chord is constructed by using notes that come from the native key of a song, then it is a **diatonic chord**. For example, in the key of C major (all the white notes on a piano C, D, E, F, G, A, B, C) the diatonic triad chords that can be naturally constructed using this scale are C, Dm, E, F, G, Am, and Bdim. There are countless examples of well-known pieces of music that exclusively rely on the use of these simple diatonic triads to construct their harmonic sequences.

However, if a chord is constructed by using one (or more) notes that do not belong to the native key of a song it is a **chromatic chord**. In the example of C major, any chord that contains at least one of the black piano notes (C#, D#, F#, G#, A#) is a chromatic chord.

Composition technique 1: The chromatic chords of *Tristram Village*

The most well-known piece of the *Diablo* franchise is the 12-string acoustic guitar piece that plays while the player is taking a break from fighting the forces of evil and pays a visit to the (seemingly) safe local village, named *Tristram*. It is interesting to note that *Tristram* is an alternate spelling for the common German name *Tristan* which might or might not have been a hidden tribute to Wagner's opera *Tristan and Isolde* and especially its use of the *Tristan* chord; a dissonant chord that resolves to another dissonant chord that has spurred endless debates among musicologists about its ambiguous harmonic functions. Mat Uelmen has not suggested any association, but the *Tristram Village* theme is full of mysterious and dissonant chords that resolve into dissonant chords. This harmonic development can be interpreted in many ways, but the point of this chapter is not to advocate a specific interpretation but rather to demonstrate the technique of using chromatic chords to introduce unexpected musical colours. The word chromatic originates from the Greek word chroma (χρώμα) which literally means colour.

Harmonic analysis of the Tristram *Village opening section*

Have a listen to the opening chords of the theme (video example 30) until 1:14" while looking at the harmonic analysis of this section in Figure 8.1. As you can probably immediately observe, none of the chords used are your typical major or minor diatonic triads. Although the piece is predominately based in A minor, there is no use of the A minor chord anywhere in this section. In fact, except for the final chord that resolves in Em there is no use of any basic minor or major triads anywhere. In their place, we find an extended use of numerous chromatic chords that are echoed in groups of two.

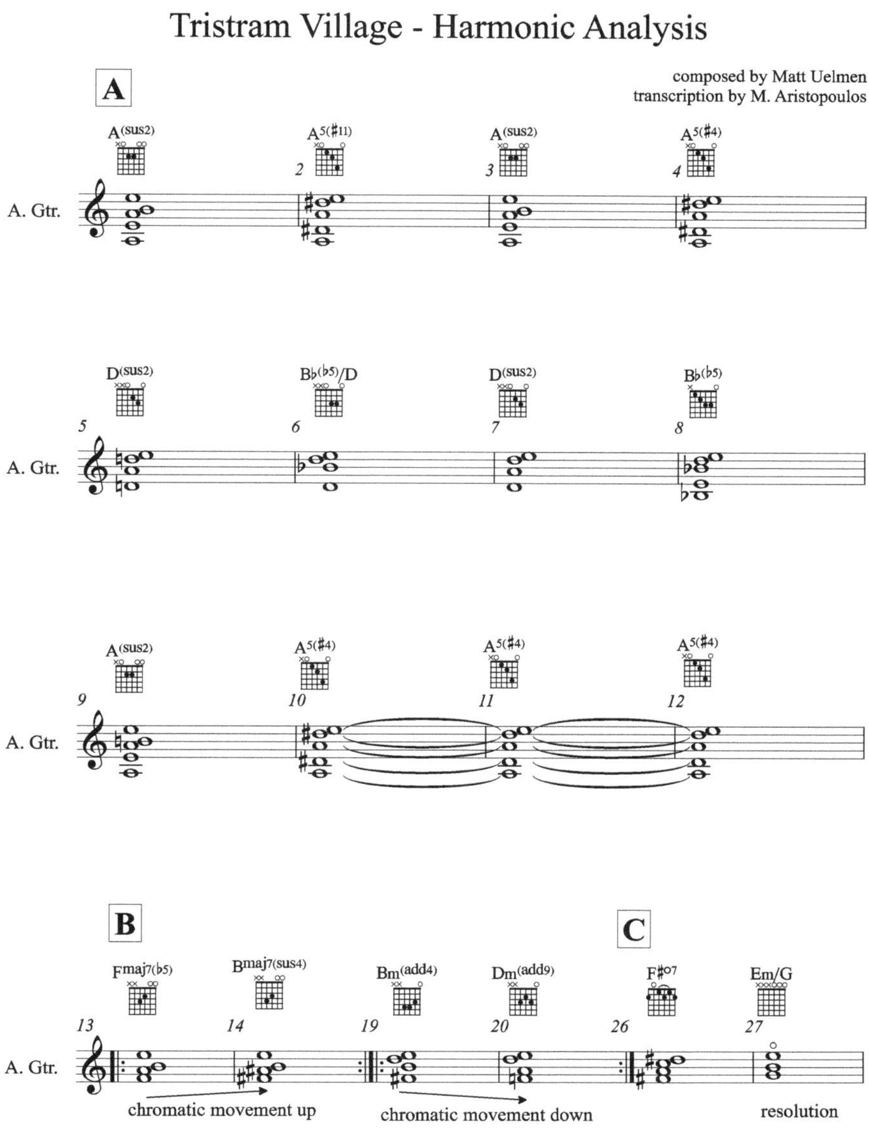

Figure 8.1 A harmonic analysis of the opening section of the *Tristram Village* theme.

The qualities of these chromatic chords are based around three different types of alterations to the diatonic triads:

1) The sus2 chord (1, 2, 5) which simply omits the third (the note that defines the major/minor quality of the chord) and replaces it with a major second.
2) The $5^{(\#4)}$ chord (1 #4 5) which omits the third and replaces it with an augmented fourth. This is an especially dissonant chord as the augmented fourth produces a tritone with the tonic as well as a minor second with the perfect fifth (ex: A, D#, E). It can also be respelled as a $5^{(\#11)}$ and although no such formal use of this type exists, it can be argued that it can be classified as a sus#4 chord.
3) The b5 chord (1 3 b5) is a major chord that has its perfect fifth flattened and thus produces a tritone with the tonic (ex: Bb, D#, E).

The chromatic chords in *Tristram* are always sequenced together in groups of two in a question-and-answer format that is always repeated at least once before the harmony develops further. In section A, the first pair of A^{sus2} – $A5^{(\#4)}$ is repeated twice before moving to the repetition of the pair D^{sus2} – Bb^{b5} and then echoed again from bars 9–12. This constant repetition of chromatic sequences helps establish the qualities of each chord before the piece progresses and achieves a sense of coherence in a harmonically ambiguous context. The melodic lines of the other instruments (not notated here) add an additional layer of harmonic complexity as they move between the A minor scale and the chromatic alterations of the chords.

In section B, these chord types are expanded further to include the use of major sevenths and the pair B $Fmaj7^{(b5)}$ – $Bmaj7^{(sus4)}$ before moving to the next chromatic pair that adds 9th and 11ths. The voice leading in section B is particularly effective as the bottom two voices move up and down chromatically while the top two voices remain unchanged using the open strings of the guitar, a technique that is regularly explored in this song.

Traditionally sus2 and sus4 chords are formed by holding a tone from a previous chord that has been suspended and needs to be resolved upwards or downwards to the third. However, none of the suspended notes resolve to the third here nor do the tritones resolve to perfect fifths. The only resolution that follows the rules of traditional classical harmony arrives in section C bar 27 (01:00") where a diminished 7th chord finally resolves fully to a minor chord and creates a sense of conclusion before the piece moves on to further harmonic adventures. It can therefore be argued that this entire highly dissonant section is similar to a prolonged version of Wagner's use of the *Tristan* chord with a prolonged resolution lasting 28 bars!

Composition technique 2: Ambient music as an open-ended storytelling device

One of the fundamental premises of *Diablo* is that it offers an almost never-ending adventure within a procedurally randomized world. As you can see in Chapter 2 – *Ballblazer*, Chapter 14 – *Apotheon*, and Chapter 15 – *No Man's Sky*, there are many generative music techniques that can be implemented to produce variation. However, the music of *Diablo* does not contain any generative elements despite the game world itself making heavy use of randomization to produce new content (see Figure 8.2). Moreover, contrary to most of the case studies explored in this book, the music is not particularly interactive either. The game consists of extended ambient themes that are designed to evoke a particular atmosphere for each major location in the game. Each theme is looped for as long as the player remains in that area and it does not adapt to the developing action in any of the usual ways: there are no boss themes, no quest triggers, no exploration/battle modes, no stingers, nor any other kind of interactive techniques commonly found in most game soundtracks.

Instead, the composer intentionally wanted the music to remain as open ended as possible in terms of its suggestive experience. Although such an approach might appear counterintuitive at first, it works remarkably well within the action heavy context of this game as players might encounter similar types of generated events thousands of times and any clearly

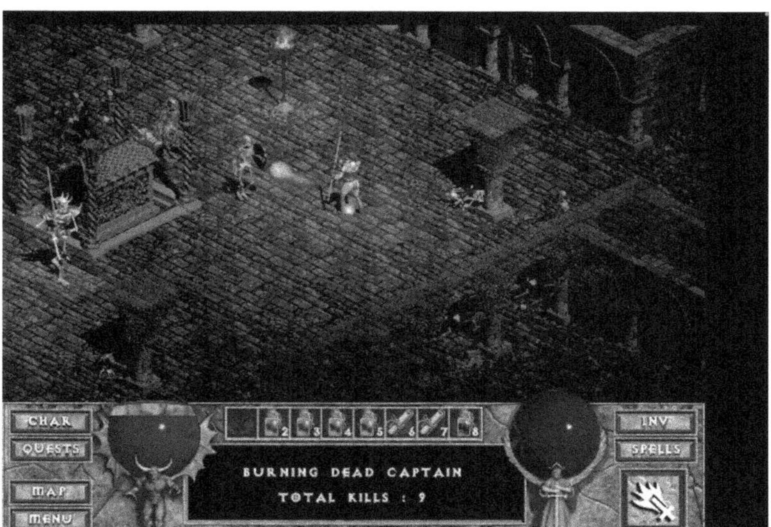

Figure 8.2 A gameplay screenshot from *Diablo* demonstrating a randomly generated dungeon design during a main quest.

distinguishable musical responses would quickly become repetitive and intrusive (ex: battle starts/ends). The overall feeling and atmosphere of each area is represented in the music, but the details of the adventure are left to the player's imagination. In an earlier interview, Matt Uelman stated that "it's kind of a piece that never really goes anywhere. It's funny – it's a hard thing to do, because every musician wants to take people on a journey"[2]. It is noteworthy that many Blizzard's games that followed *Diablo* often have a similar non adaptive and rather ambient approach but, in my opinion, none have done so as successfully, precisely because of the use of a non-functional musical language that leaves space for multiple interpretations. Perhaps by avoiding telling a very specific story, the composer can also avoid the risk of telling one that is different to the one that is being played out.

Most other level tracks of the soundtrack are harmonically simpler than *Tristram Village,* but they similarly use an ambiguous tonality and harmonic direction. This ambient aesthetic approach, aside from being highly suitable to the dark and haunting atmosphere of the game, has some additional advantages that can be beneficial to a game composer's toolbox. First, the lack of a predictable harmonic direction assists in concealing the position and length of the loop as the listener cannot easily identify its position. This makes the music feel more seamless and less repetitive which is especially useful considering the soundtrack's relatively short duration. Second, any sudden transitions to other musical segments that are occasionally triggered by sudden developments in the gameplay (ex: level change, player death, menu) often might feel less jarring as the harmonic journey is not interrupted in the same way as in a composition based on functional harmonic development.

Production tools – Lo-fi charm

The production aesthetic of *Diablo* has a distinct sound that does not match the crispiness found in other 16-bit soundtracks of the same era. According to Matt Uelman the music production of the soundtrack was very low budget. His setup was built around an *Ensoniq ASR-10* hardware sampler, an entry level *AKG* microphone, and a primitive DAW of the time called *Sound forge.* He states:

> I think Diablo accidentally had extra originality to it just because my whole approach was so low budget, structuring everything around that ASR-10, it gave me a distinctive sound compared to what someone much more pro would have been doing with the standard LA soundtrack hack Akai libraries of the time.[3]

The *ASR-10* shipped with 2MB of internal memory which was expandable to 16MB and came with a number of floppy disks containing stock samples.

It was a complete production studio in a box featuring a sequencer, a recorder sampler capable of capturing 30 kHz or 44.1 kHz rates at 16 bit, and 62 types of effects processing including a vocoder. The sampler is it is still popular today and it frequently sells for approximately over \$1K[4] because of its idiosyncratic sound and unusual architecture that allows eight simultaneous sample layers that can be modulated by numerous synth envelopes.

Matt made extensive use of the effects processing of the *ASR-10* with frequent use of echo as a rhythmic effect, prolonged reverbs, and constant pitch shifting. The combination of acoustic instruments with processed natural sounds come from psychedelic rock bands of the 1960s and 1970s such as *Pink Floyd*, *Led Zeppelin*, and *Bauhaus*. However, this approach is taken a few steps further in *Diablo* in order to create the demonic and hellish soundscape that the game required. You can regularly find processed and disturbing recordings of human agony that are discretely mixed with the rest of the music: screaming, breathing, grunts of pain, laughter and even distant baby cries. There are also sampled elements borrowed from church music such as bells, angelic choirs, organs, and voices that add to the gothic atmosphere (video example 31).

Another important technique that contributed to the distinct sound of the music, was that all the files were down sampled to 22,050 Hz to accommodate the Windows 32 release, and even further to 11,025 Hz for the PlayStation 1 release! Down sampling has no impact on the playback speed or pitch of the signal but according to the *Nyquist-Shannon theorem* your sampling rate needs to be at least double than the highest sampled frequency to avoid a sampling error known as aliasing. For example, a sampling rate of 22,050 Hz will only accurately represent frequencies up to 11,025 Hz. Therefore, the low sampling rate of the natural sounds and instruments in the music caused a lot of the higher frequencies to be less accurate and the overall production to become more unclear. This low fidelity sampling matches well together with the low resolution of the visuals and is a technique that can easily be recreated to evoke a sense of 1990s nostalgia in contemporary games (see task 2).

Takeaway tasks

These two tasks can be combined.

Task 1 – Composition (challenging) – Compose a theme for an area in Diablo *that makes use of chromatic chords*

After you feel comfortable with using basic diatonic triads the next step is to start exploring chromatic chords. You can begin by writing a diatonic chord sequence and then experiment altering one or more notes of each chord. To achieve a sense of coherency, try to rely only on a few chord structures that are appealing to you and avoid making a sequence that incorporates too many new chord types at the same time unless you want the piece to feel completely atonal. There are plenty of harmonic possibilities to explore here: aug, b5, sus2, sus4, borrowed major/minor/dim chords from other keys. You can also go beyond triads and explore chromatic extended chords where the possibilities are much greater but so is the harmonic complexity (ex: add 9ths, 11ths, 13ths).

Task 2 – Production (easy) – Emulate the sound of early digital samplers

A straightforward way of emulating the limited capabilities and sonic characteristics of older pieces of digital recording and sampling gear such as those explored in *Diablo* is to reduce the bit depth and sample rates of your audio files. Most DAWs come with a plug-in effect that can be used it for this purpose: *Logic Pro* and *Cubase* come with *Bitcrusher* (Figure 8.3), *Ableton Live* with *Redux*, and *Pro Tools* with the *D-fi* family of plugins.

Using *Logic*'s *Bitcrusher* you can change the down sampling knob from the value of 1x, that is having no effect on the signal, to a higher multiple that reduces the signal proportionally (ex: 10x reduces the sample rate to one-tenth of the original so a 48 kHz will be resampled as a 4.8 kHz). The other audio fidelity parameter is the bit resolution that usually ranges from 1 to 24 bits. Reducing the bit rate alters the precision of the sampling process and lower values will generate more distortion, introduce noise, and further sampling errors. You can experiment with these parameters and use them creatively but try to conduct before and after comparisons to begin to familiarize yourself with how sampling parameters affect different sound material. Usually, any reductions in the bit rate below 8 will introduce

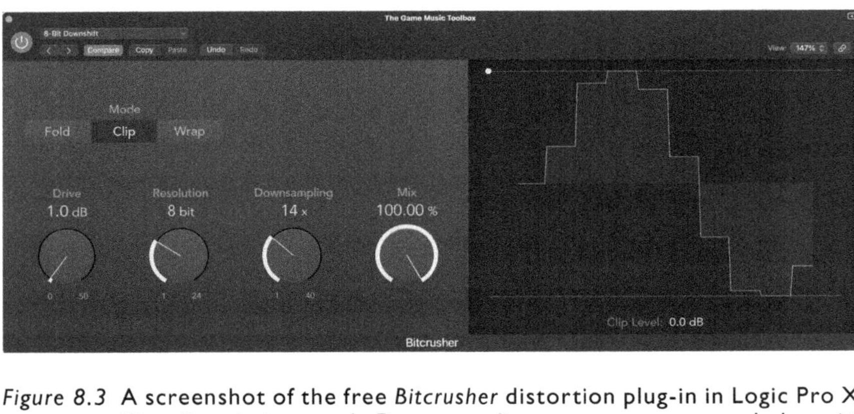

Figure 8.3 A screenshot of the free *Bitcrusher* distortion plug-in in Logic Pro X. The Resolution and Downsampling parameters can help with reducing the audio fidelity of your sounds.

significant distortion as they raise the noise level of the recording. Reductions in sampling rate behave quite differently depending on the harmonics of your recorded material and you can get some interesting coloration in your higher frequencies that can be used creatively.

Notes

1 "Matt Uelmen - The Music Of Diablo 1996 - 2011: Diablo 15 Year Anniversary."
2 Breckon and Uelmen, "From Tristram To Torchlight: An Interview With Composer Matt Uelmen."
3 Uelmen, "Interview With Matt Uelmen."
4 "Ensoniq ASR-10 |."

Bibliography

Breckon, Nick, and Matt Uelmen. "From Tristram to Torchlight: An Interview with Composer Matt Uelmen". *Shacknews*, 2009. https://www.shacknews.com/article/60997/from-tristram-to-torchlight-an.

"Ensoniq ASR-10 |". *Vintage Synth Explorer*. Accessed 1 October 2022. https://www.vintagesynth.com/ensoniq/asr10.php.

"Matt Uelmen – The Music of Diablo 1996 – 2011: Diablo 15 Year Anniversary". *Discogs*. Accessed 1 October 2022. https://www.discogs.com/Matt-Uelmen-The-Music-Of-Diablo-1996-2011-Diablo-15-Year-Anniversary/release/3243711.

Uelmen, Matt. "Interview with Matt Uelmen". *Games Today*, 2019. https://gamestoday.info/pc/diablo/interview-with-matt-uelmen/.

Chapter 9

Assassin's Creed

Music as a time travelling device in four historical games of the franchise

About the games

Assassin's Creed is an open-world action-adventure video game franchise developed by Ubisoft. It follows a variety of stealthy fictional assassins in their quest to restore peace within different historical settings.

Fun facts

Many *Assassin's Creed* games, feature Discovery Tours "that let visitors freely roam Ancient Greece, Ancient Egypt and the Viking Age to learn more about their history and daily life. Students, teachers, non-gamers, and players can discover these eras at their own pace, or embark on guided tours and stories curated by historians and experts."[1]

How did the composers get the gig?

There are multiple composers involved in these games, each with their own specialties and personal career journey. Jesper Kyd, is one of the most established and busy game composers of AAA games, he wrote the original scores for the first three games of the franchise and returned to co-write the music for *AC Valhalla*. Sarah Schachner, after graduating from Berklee College of music began assisting Hollywood composer Brian Tyler who wrote the music for *AC Black Flag*. She then got a headline composing role for *AC Origins*, and *AC Valhalla*. *AC Syndicate* was written by Austin Wintory that has had a flourishing career since his work on Journey came to prominence (see Chapter 10). Einar Selvik, had previously worked on the TV series Vikings before being hired to work on *AC Valhalla*. He is a Norse music specialist that frequently lectures on historical Norse music as well as performs as the lead singer with his well-known folk band Wardruna.

DOI: 10.4324/9781003146872-10

Historical authenticity 101

Music can be a very powerful tool for evoking various historical settings and cultures. For example, listening to a few seconds of a sliding electric guitar accompanied by a percussive rattle might easily conjure a setting in the Wild West, while a pentatonic melody on a bamboo flute accompanied by taiko drums might be suggestive of a Samurai setting. Musical reimaginations of historical periods often contain stereotypes and inaccuracies. For instance, even though the electric guitar is frequently present in many Wild West themed soundtracks (ex: in the *Red Dead Redemption* franchise), the instrument had clearly not been invented until after the era known as the American frontier (1607–1912). The connection of the electric guitar to this context originated from the popular Spaghetti Western films of the 1970s and the music of legendary composer Ennio Morricone. Authenticity is sacrificed here in favour of entertainment as the use of the guitar effectively relates to other cultural associations that the audience might have of that era: the presence of nylon stringed guitars, the discovery of electricity, the railroad, and the rugged atmosphere of cowboy shootouts.

Game composers usually reimagine historical music with varying degrees of authenticity in relation to the original musical cultures. Sometimes, composers might simply lack sufficient musicological knowledge, or historical precision might be sacrificed in favour of entertainment and storytelling purposes. This semi accurate historical approach often mirrors the use of creative freedoms in historical game design that tend to prioritize excitement and fun (Figure 9.1). Composer and musician Einar Selvik points out in an interview about his work on *AC Valhalla* that few players would actually want to play a game that fully simulates the life of an average Viking based on scientific evidence as 99.9% of the time would need to be spent on mundane tasks such as farming or fishing, and maybe a fraction of that time, if any at all, in epic raids for glory in foreign mysterious lands.[2]

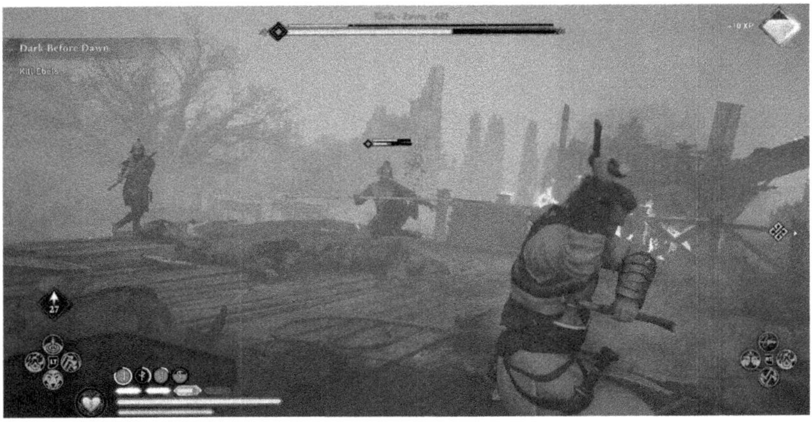

Figure 9.1 A gameplay screenshot from *Assassin's Creed Valhalla* depicting the 845 AD Viking siege of Paris. While many of the main events and locations are historically accurate, the battles are obviously filled with a dose of historical fantasy!

Composition technique 1 – Music as a time travelling device

Let us examine some of the techniques and production tools that composers used in four games of the Assassins Creed franchise to evoke four historical eras of the past.

Assassin's Creed – Origins

Period: Ancient Egypt – end of the Ptolemaic Period (49–43 BC)

Travelling back to the world of ancient Egypt, our knowledge of a culture that existed over 2,000 years ago is limited and is only based on assumptions deriving from archaeological findings. This gave composer Sarah Schachner the liberty to reimagine the music with few constrains: "I wanted to create a hybrid sound of old and new with an air of ambiguity and mystery to represent this otherworldly culture that was so immersed in mythology."[3]

Instrumentation

In terms of instrumentation Sarah relied on traditional middle eastern instruments such as the oud, lutes, lyres, bells, winds, and hand drums that were heavily processed and juxtaposed against a synth foundation. As the game also included sci-fi elements in its story, the combination of modern and traditional instruments was a logical conceptual approach.

THE HARMONIC SCALE

In many films and games the harmonic scale is one of the core elements that is used to evoke an ancient Egyptian setting, as it has a strong connotation to the Middle East (ex: the main theme from the Hollywood film *The Mummy*). Similarly, both the harmonic and the double harmonic scales are dominating many of the melodies you will hear in the *Origins* soundtrack (usually in D minor) such as in video examples 32 (from 0:44) and 33. These scales are easy to identify as they both have a very distinguishable sound that cannot be mistaken with any other scale. Their characteristic sound is produced using a three-semitone jump that is both preceded and followed by a semitone movement. The harmonic minor is the simplest variant of this scale, which contains this jump only once, instead of the double harmonic minor that has two of them in succession. The formulas to construct these two scales are as follows:

For the harmonic minor take any natural minor scale and simply raise the 7th by one semitone. For example, A minor that contains only the white notes on a piano (A, B, C, D, E, F, G, A) will become A minor harmonic by only raising the G by one semitone: A, B, C, D, E, F, **G#**, A. Likewise, D minor (D, E, F, G, A, Bb, C, D) will become D harmonic minor by raising its 7th: D, E, F, G, A, Bb, **C#**, D.

For the double harmonic minor take any minor scale, **flatten the 2nd**, and **raise the 3rd and 7th** notes by one semitone. For example, A minor double harmonic: A **Bb C#** D E F **G#** A, and D minor double harmonic: D **Eb F#** G A Bb **C#** D.

Production

Sarah relied heavily on the use of reverberation, an aesthetic that is commonly found on many soundtracks that seek to reflect ancient themes, perhaps because it can add an ethereal and mystical quality to the music. She resampled many of the instruments through a simple app on her iPad and run them through the Strymon Big Sky reverb pedal. The recurring eerie drone that is found across the soundtrack was also created by modulating high amounts of reverb feedback through her Eurorack modular synth.[4]

Assassin's Creed – Valhalla

Period: Viking expansion into Britain (9th century).

Instrumentation

To transport players into the brutal and heroic world of the Vikings, composers Jesper Kyd, Sarah Schachner, and Einar Selvik explored a collection of ancient instrument replicas, some of them directly related to the Viking culture and others chosen primarily for their timbre. The restraints of the ancient instrumentation posed some creative limitations requiring the composers to keep inventing new ways to compose. "Primitive styles of regional folk music that aren't necessarily harmonically complex can seem deceivingly easy to write and produce," says Sarah Schachner. "But that was far from the case. The rustic instruments are fairly limited and were not easy to play. I was continuously trying to find new ways to write for them to keep the score from feeling repetitive."

Here is a selection of some of the most interesting historical instruments used in the soundtrack:

Morin Khuur – Is a Mongolian two stringed lute also known as a "horse head fiddle". The name comes from the Mongolian legends surrounding the instrument's origins in which the hair of beloved horses were used after their death to create the strings and bow of the instrument.[5] It was one of Jesper Kyd's favourite instruments in the soundtrack.[6]

Carnyx – A Celtic war horn made from bronze that was used in battle. The pitch is manipulated only by a change in breath and embouchure. For *AC Valhalla*, Schachner chose the Deskford version of the Carynx because of its raspy tones an unusual harmonic series which directly influenced the compositional approach of the pieces it was used on.[7]

Skalmejen – A type of primitive oboe capable of basic non chromatic melodic lines.

Lyre – A instrument dating back to 1400 BC. The iteration thought to have been used in the Viking age was a 7-string Nordic lyre tuned to the Pentatonic scale. According to archaeological findings this was the most common instrument in the Nordic regions at that time.[8]

Tagelharpa – A bowed version of the Nordic lyre (see Figure 9.2). The instrument allows the player to maintain a root note drone and play melody lines on top and was used extensively in the *Valhalla* soundtrack.

Paleolithic Flutes (Bone Flutes) – A type of primitive flute or recorder made from hollowed animal bones. In the Viking age, these were often made from sheep bones.[9]

Animal Hyde Frame Drum – A small to medium sized hand drum often made with Goat Skin, played with a variety of beaters, or with the palm of your hand.

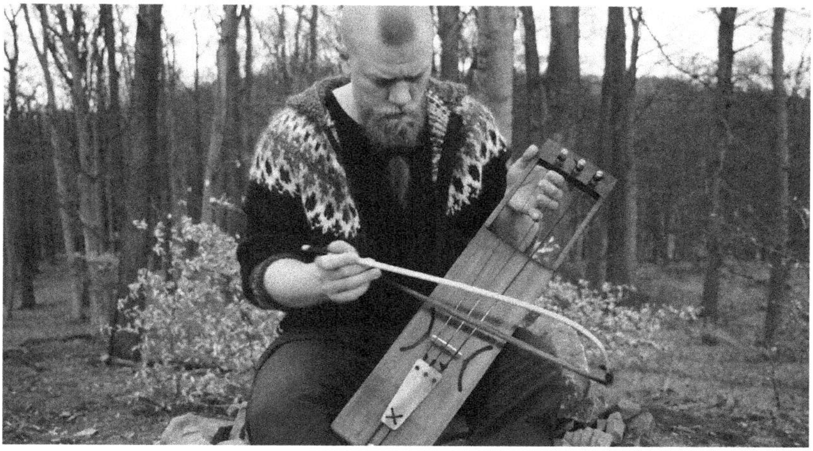

Figure 9.2 A photo of musician playing a Tagelharpa similar to the ones by Einar in *AC Valhalla*. Photo provided by musician and instrument maker VeduvianArt.[10]

Viking vocal techniques

According to an early account of Viking singing by Roman Emperor Julian Apostata in 350 AD, the singing of Vikings sounded similar to the cries of crows.[11] This could be a reference to **Kulining** singing: a Scandinavian high pitched vocal call that was used to communicate with animals over great distances. Other historical accounts describe Vikings having a raspy and throat-centric type of singing possibly originating from the heavy drinking that took place in war celebrations. Einar Selvik's voice played a fundamental role in conjuring the Viking spirit in the game. You can listen to his unique vocal interpretations of Viking singing throughout the soundtrack and in particular in video example 34 Odin's Ride to Hel and video example 35 (from 01:45") Skullcrusher.

One of the pieces that Einar considered to be the closest to an authentic reimaginations of Viking music in the game is Lust for Battle (video example 36). The song is based on the Skaldic tradition which is the Viking version of an Irish bard. According to Einar: "The Norse culture was predominantly an oral society and so we clearly see that in the oldest song traditions we have here in the north, rhythms and melody are often guided by the (often) complex poetic structures."[12] The lyrics are an excerpt of an old Viking poem which expresses the rousing and build up before a battle, and it is sung by the in the game characters when you travel with your warship in anticipation of the next raid. The rhythm and the melody are guided by the poem, and is cyclical in nature, with multiple cycles of two melodic phrases around C natural minor: (1) C-F-Eb-Eb-D x4, and (2) a melisma of C-F-Eb-F-G-Ab-G-C. **Melisma** is a simple but ancient vocal technique that can be traced all the way to the Greek Eleusinian Mysteries, in which the same syllable is held over multiple note runs. It is interesting to note that the 7th tone of the minor scale is never used in the song perhaps suggesting a peculiar pentatonic (five notes) or hexatonic (six tones) tunning partly based in C natural minor.

Production

Evoking a historical atmosphere can also be explored through the use of different recording techniques. Sarah Schachner recorded all the stringed instruments with a very close-miked setup to emphasize their raw and imperfect characteristics. She used a large diaphragm condenser Neumann TLM 103 running through a Manley Force 4 track tube preamp. Many of the sounds were run through external processing such as the Elektron Analog heat distortion/saturation effects unit, and the Strymon Big Sky reverb."[13] On the contrary, Jesper Kyd chose to record using a long-distance miking setup: "I would record things and the mics would be quite far from the instrument; I would have this air in the recording to simulate being outside amongst the mountains, fjords and forests".

Assassin's Creed II

Period: *Italian Renaissance (end of 15th century)*

The Italian Renaissance is a historical setting in which we have a very clear idea of how music sounded like, contrary to the earlier examples of ancient Egyptian and Viking music. However, Ubisoft did not want Jesper Kyd to compose very realistic renaissance music as they thought it might feel too boring for a modern gaming audience.[14] Instead, Jesper chose to highlight the emotional aspects of the dramatic story focusing on the main character Ezio, and his family.

Ezio's Family (video example 37) is arguably the most famous piece from the *Assassin's Creed* universe but also one of the simplest. This track became so iconic that it was eventually used in many sequel games that followed even though they had nothing to do with that character. Jesper said the following about it:

> I originally envisioned the theme to represent Ezio's loss and struggles and I tried to capture the emotion Ezio felt when thinking about this act of betrayal and tragedy. This moment defines who he is and who he becomes and of course why he joins the brotherhood and becomes an

> Assassin. There is always a sacrifice and struggle for all the characters in the series when they join this secret brotherhood and to me, that's what Ezio's Family has come to represent. It's absolutely wonderful to hear all the different versions of Ezio's Family not only in the games but also the many fan versions on YouTube. This theme has evolved so far beyond anything I could have imagined.[15]

Even though the music that Jesper wrote is clearly not sounding anything like authentic renaissance music, there are musical elements that still contain hints of the rich musical culture of the era such as the use of lute-type instruments, the operatic use of voice, the dramatic character of the music, and most prominently the central use of **ostinato:** a repeating short phrase that forms the basis for the entire composition. Ostinato is the Italian word for "stubborn" and it was frequently used by renaissance composers as a basis for variation, such as in the operas and sacred works of Claudio Monteverdi. The *Ezio's Family* theme is based entirely on a minimalistic **ostinato** of only two bars long consisting of eight continuous quarter notes in a question-and-answer format between each bar. This short melody is memorable and easy to sing, but it is orchestrated beautifully by moving the ostinato voice across different instruments in the arrangement which creates a natural sense of development despite the melody and chord progression of Dm-C-Bb-Dm remaining unchanged. Listen to the piece (video example 37) while observing how the ostinato moves between approximately 20 different instrumentation variations.

Assassin's Creed – Syndicate

Period: *Victorian London during the 2nd Industrial Revolution (mid-19th century)*

AC Syndicate is set in 1868 Victorian London during the second industrial revolution. It focuses on the political inequalities between the industrial workers and their struggle for liberation from the Templar cult. It is interesting to note, that similarly to the Italian Renaissance setting of AC II, composer Austin Wintory did not place historical realism as the highest priority of his score:

> I think that a score should be primarily focused on the characters and of the ideas being presented by a game and trying to figure out the subtext of those. If it manages to wink towards the era while doing those things then great. But I don't think it's necessary to build the score around the time period as the starting position because the game already gives you that.[16]

One musical element of Austin's music that clearly connects it to Victorian London was his invention of "murder ballads" that are sung by in-game characters during pivotal story moments, as well as by in-game musicians in various London locations such as pubs or street alleys. The songs are based on a type of theatrical entertainment that was popular in London at the time known as the Music Hall genre. Have a listen to the murder ballads in video examples 38–40 while looking at the analysis in Table 9.1. The songs share many similar characteristics that were prominent in this genre: they rely on a solo vocal and piano accompaniment, repetitive lyrics, and catchy melodies. The harmonic direction is simple and cyclical, always starting on the tonic (I), usually arriving at the dominant (V) half-way through the phrase, and then returning to the tonic via the use of a common cadence such as II-V-I or IV-V-I. These features were deliberately implemented in the music to allow the (usually drunk) audiences of music halls to quickly learn the tunes and easily sing along. The introduction of the music hall genre by Wintory was a clever method of adding a historical touch to some of the music in the game without limiting the language of the rest of the score.

Table 9.1 Harmonic analysis of three diegetic songs from A.C. syndicate

'Give Me the Cure' – Austin Wintory
Piano & Voice – 3/8 – ♩. = 70 – Verse Sequence

Bar: 1	*2*	*3*	*4*	*5*	*6*	*7*	*8*
I	I	vi_{sus4}	vi	IV	IV	V	V
Bar: 9	*10*	*11*	*12*	*13*	*14*	*15*	*16*
V	V	vi	I	IV	V	I	I

'The Late Pearl Attaway' – Austin Wintory
Piano & Voice – 4/4 – ♩ = 120 – Verse Sequence

Bar: 1	*2*	*3*	*4*
I	IV	I	Vsus4|V
I	IV|vi	ii|V	I

'Feating on a Lord' – Austin Wintory
Piano & Voice – 4/4 – ♩ = 95

Bar: 1	*2*	*3*	*4*
i|V	i	vi	I
III|vi	vi|V	i	I

Takeaway tasks

Task 1 – Composition (difficulty depends on the chosen period) – Compose a theme that is evocative of a specific historical setting

You can select a historical setting from any of the 20 AC franchise games in existence. Feel free to use any of the scales, instruments, forms, and recording/production techniques discussed in this chapter, or even conduct your own research to other relevant musical culture characteristics. You can test the success of this task by doing a simple blind test with friends/family. Can they identify the time period only by listening to the music? Remember, the success might also depend on the perception of a listener with a particular culture, but the average audience member is already exposed to a wide range of period music through Netflix, games, adverts, and other media.

A note of caution for your future work with historical games; extracting only selected elements from other musical cultures especially in cases in which there are unbalanced power dynamics, might be considered as cultural appropriation if it is executed poorly and without consideration. This is an ethical question that in my opinion composers should pause to reflect upon before replicating any harmful stereotypes, as the commercial power of games can exert a strong influence on cultural perceptions.

Notes

1 "Discovery Tour by Ubisoft: Teacher Learning Resources."
2 Selvik, *Einar Selvik on Assassin's Creed Valhalla.*
3 Parisi and Schachner, "Sarah Schachner Sojourns to Ancient Egypt for 'Assassin's Creed Origins' Score."
4 Reseigh-Lincoln and Schachner, "Composer Sarah Schachner on Bringing Ancient Egypt to Life in Her Assassin's Creed Origins Soundtrack."
5 "Mongolia – Morin Khuur (Horse Head Fiddle)."
6 Mesecher and Kyd, "Q&A with Assassin's Creed Valhalla Composer: Jesper Kyd."
7 Mesecher and Schachner, "Q&A with Assassin's Creed Valhalla Composer: Sarah Schachner."
8 Pope and Selvik, "Interview: Einar Selvik Talks Assassin's Creed Valhalla."
9 "Historic Pastimes & Musical Instruments."
10 VeduvianArt, *Selfmade Tagelharpa – Jouhikko ANNAÐ (Forest Clearing).*
11 Friis, "Vikings and Music."
12 Pope and Selvik, "Interview: Einar Selvik Talks Assassin's Creed Valhalla."
13 Sundstrom and Schachner, "History in the Making: Scoring Assassins' Creed Valhalla."
14 Price and Kyd, "Jesper Kyd Interview – Revelations."
15 Blackett and Kyd, "Jesper Kyd Interview."
16 Ebbinghaus and Wintory, "Composer Austin Wintory about Scoring Assassin's Creed Syndicate."

Bibliography

Blackett, Colum, and Jesper Kyd. "Jesper Kyd Interview". *The Ones Who Came Before*, 2019. https://www.theoneswhocamebefore.com/jesper-kyd-interview.

"Discovery Tour By Ubisoft: Teacher Learning Resources". *Ubisoft*. Accessed 2 October 2022. https://www.ubisoft.com/en-gb/game/assassins-creed/discovery-tour.

Ebbinghaus, Peter F., and Austin Wintory. "Composer Austin Wintory about Scoring Assassin's Creed Syndicate". *Behind the Audio*, 2015. https://behindtheaudio.com/2015/12/composer-austin-wintory-about-scoring-assassins-creed-syndicate/.

Friis, Mogens. "Vikings and Music". *Viking.No*, 2004. https://www.viking.no/s/life/music/d-musikk-mogens.html.

"Historic Pastimes & Musical Instruments". *York Archaeology*, 2019. https://www.yorkarchaeology.co.uk/resilience-year-2/2019/6/17/historic-pastimes-amp-musical-instruments.

Mesecher, Andy, and Jesper Kyd. "Q&A with Assassin's Creed Valhalla Composer: Jesper Kyd". *Music Connection Magazine*, 2020. https://www.musicconnection.com/qa-with-assassins-creed-valhalla-composer-jesper-kyd/.

Mesecher, Andy, and Sarah Schachner. "Q&A with Assassin's Creed Valhalla Composer: Sarah Schachner". *Music Connection Magazine*, 2020. https://www.musicconnection.com/qa-with-assassins-creed-composer-sarah-schachner/.

"Mongolia – Morin Khuur (Horse Head Fiddle)". *Royal Collection Trust*. Accessed 2 October 2022. https://www.rct.uk/collection/95705/morin-khuur-horse-head-fiddle.

Parisi, Paula, and Sarah Schachner. "Sarah Schachner Sojourns to Ancient Egypt for 'Assassin'S Creed Origins' Score". *Billboard.Com*, 2017. https://www.billboard.com/music/music-news/sarah-schachner-ancient-egypt-assassins-creed-origins-8005935/.

Pope, Erica, and Einar Selvik. "Interview: Einar Selvik Talks Assassin'S Creed Valhalla". *Soundtracks, Scores And More!*, 2020. https://soundtracksscoresandmore.com/2020/11/13/interveiw-einar-selvik-talks-assassins-creed-valhalla/.

Price, Andy, and Jesper Kyd. "Jesper Kyd Interview – Revelations". *Music Tech*, 2018. https://musictech.com/features/jesper-kyd-interview-revelations/.

Reseigh-Lincoln, Dom, and Sarah Schachner. "Composer Sarah Schachner on Bringing Ancient Egypt to Life in Her Assassin'S Creed Origins Soundtrack". *Music Radar*, 2018. https://www.musicradar.com/news/composer-sarah-schachner-on-bringing-ancient-egypt-to-life-in-her-assassins-creed-origins-soundtrack.

Selvik, Einar. *Einar Selvik on Assassin's Creed Valhalla*. Video, 2021. https://www.youtube.com/watch?v=69JKD5Ic85Q&ab_channel=FaceCulture.

Sundstrom, Matthias, and Sarah Schachner. "History in the Making: Scoring Assassins' Creed Valhalla". *Music Tech*, 2020. https://musictech.com/features/scoring-assassins-creed-valhalla-history-in-the-making/.

VeduvianArt. *Selfmade Tagelharpa – Jouhikko ANNAÐ (Forest Clearing)*. Video, 2021. https://www.youtube.com/watch?v=8GehySR5nMw&ab_channel=VedunianArt.

Chapter 10

Journey (2012)

A masterclass in monothematic scoring

About the game

Journey is a third person story adventure game in which you take on the role of a silent protagonist that meditatively wanders across a vast desert. The player discovers the ruins of what was once a thriving civilization and will have to battle alone, or with the company of other online players, against sandstorms, snow, and wind in a transformational journey towards the peak of a distant mountain.

Fun facts

The music takes the auditory and narrative lead throughout the game as there is minimal sound design and no dialogue. It made history by being the first video game to receive a GRAMMY nomination for its soundtrack.[1]

How did the composer get the gig?

During his studies at USC, composer Austin Wintory wrote the music for a game called *Flow* that was another student's Master's thesis. This indie game unexpectedly went viral, and Sony picked it up for full-scale production as a PS3 title.[2] Austin stayed in contact with the original team, and eventually collaborated again in *Journey* for the full three years of its development.[3] The success of this soundtrack was pivotal for Austin's career[4] and helped establish him as one of the most celebrated game composers (see his work on *Assassin's Creed* in Chapter 9).

Composition technique – Monothematic scoring

This soundtrack is a masterclass in monothematic scoring, the art of relying on a single central theme for the entire soundtrack. The theme is used as a symbol of you, the player, and develops in numerous interesting as the journey unfolds. Let us examine some of the techniques Wintory uses

DOI: 10.4324/9781003146872-11

to develop his main theme by comparing three different musical moments taken from three different points of the journey: the beginning (Nascene), the middle (Threshold), and the end (Apotheosis) (Figure 10.1).

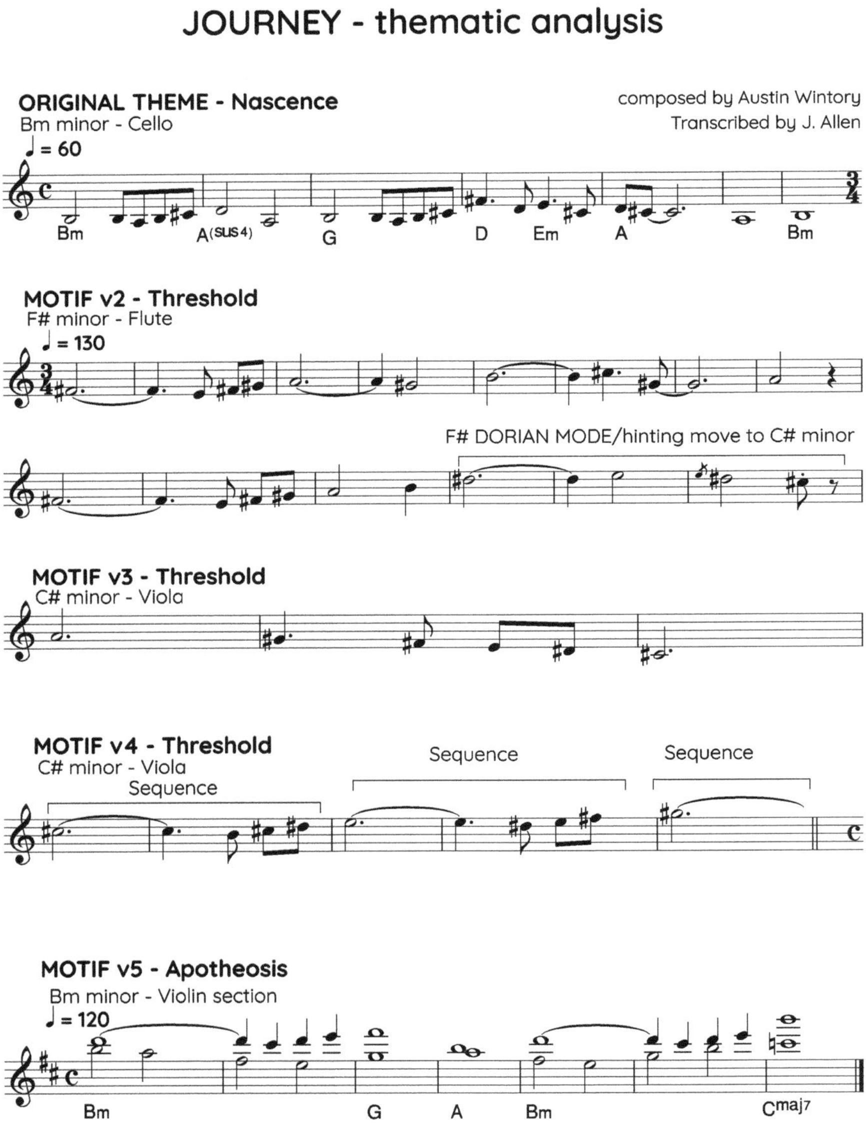

Figure 10.1 An analysis of the thematic development of the original *Journey* theme at different moments in the game. All notes have been placed in the G clef for simplicity.

Nascence

Austin wrote this theme (video example 41) the day he was hired months before any game development even begun. Surprisingly it does not appear in this version anywhere in the game, but it forms the basis of every single note he wrote for the rest of the soundtrack. The theme is introduced as an 8-bar melody played on a solo cello (performed by Tina Guo), with no underlying harmony, an approach that leaves space for different harmonic interpretations. It is then repeated an octave above on the bass flute accompanied by a pizzicato double bass line and harp chords. The theme feels emotionally ambiguous, neither happy nor sad. Although the melody is strongly centred around the notes of the Bm chord, Austin evades a firm establishment of the theme in the Bm minor tonality by avoiding the use of the F# dominant chord and hinting towards the parallel major of D, as he uses the I, IV, V chords (D, G, A). The theme is then played a third time, but the size of the ensemble grows considerably with the C flute doubling the cello in a high register while a descending counter melody is played by a large string orchestra, perhaps to foreshadow that this is going to be an epic and emotional journey.

There are traces of musical influences that can point towards different cultures, but the overall musical language is difficult to place within a specific context. This approach was something that Austin carefully planned. He recalls:

> I didn't want the music to feel ethnic or cultural in any way. The civilization around you in the game has influences in various societies but we really wanted the music to feel timeless and universal. So I didn't make an effort to draw from anywhere, and in fact a few times the music accidentally sounded like it was from various cultures and so I would change to get rid of it. Like an early bit of music had some percussion which almost made it sound Arabic so we took that out. Or at one point it felt a little Irish so I change that too, etc.[5]

Threshold

This track (video example 42) encompasses most of the music you encounter in the open desert after first stumbling on to the desert creatures.[6] At this point in the journey, the arrangement is still based on a small ensemble and has not yet built into a full orchestra. An interesting point is that the harp and viola are only playing when another player joins your journey and their level in the mix depends on the distance between you (Figure 10.2).

A lot of the thematic material is clearly recognizable from the original theme (Figure 10.1), but some parts have started to transform. The meter has shifted from 4/4 to 3/4 and we have a slightly faster tempo of 130 bpm that gives the melody a lighter and more playful feel. In this piece we can

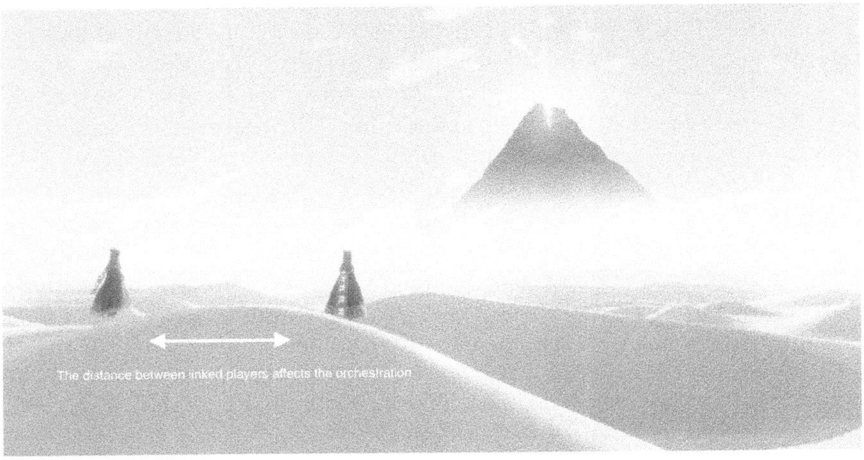

Figure 10.2 A gameplay screenshot from *Journey* showing two players playing the game as a co-op. These connections happen randomly, and players can choose to travel together or continue alone at any moment. The usernames of everyone you met across the way are displayed at the end of the game.

observe some simple but useful thematic development techniques that Wintory implements to develop the thematic material. He often keeps the first half of the melody close to its original form to maintain thematic consistency but then develops the second half to new unexplored directions. We hear these variations multiple times in the piece, which suggests that they are intentional theme transformations, rather than free flowing material.

- Motif v2 (video example 42, 0:41) is similar to the original motif but the melody is now centered around F# minor in the flute and in the second half it temporarily travels to F# Dorian minor that has a distinctive quality. This is achieved by raising the 6th tone from D to D# (for more info on modes see Chapter 7 – *Mortal Kombat*).
- Motif v3 (video example 42, 1:17) follows the same rhythm as the original theme; however, the melodic line, now in C# minor in the viola, is inverted leaving us with a descending motion downwards. By this point, the rhythm has become familiar enough for the inversion to resemble the theme even though the melodic direction is different.
- Motif v4 (video example 42, 1: 32) is almost identical to the first bar of the original aside from the placement in C# minor tonality in the viola, but in this instance rather than continuing the development of melody with the familiar melodic jump after the first bar, we get an exact repetition of the same motif but starting a note higher, a technique known as a tonal sequence (for more info on sequences see Chapter 3 – *Zelda*).

Apotheosis

This version of the theme (video example 43) plays during the final push to the summit of the mountain. The melody is very similar to the original theme with some small yet powerful differences. The arrangement has grown to include a full string section which increasingly moves higher in register as the players continues their ascends above the clouds and towards their apotheosis (an ancient Greek word meaning an elevation to a divine level). Wintory brings the theme back to its roots in Motif v5 (video example 43, 2:54, Figure 10.1) by applying the following powerful techniques:

1) **Rhythmic augmentation and diminution**
   The tempo has changed from 70 in the opening theme to a faster pace of 120 in this piece to accommodate for a more energetic rhythmic accompaniment and the metre has returned to the original 4/4. However, the pace of the melody feels slower and more lyrical at the same time. To achieve this, Wintory uses a technique known as a **perfect rhythmic augmentation** to expand the rhythmic values of the melody by an equal amount, therefore slowing it down by half without changing the tempo. The opposite technique is a rhythmic diminution in which the duration of each note would be cut in half. Motif v5 is nearly identical to the original, but the half notes have become whole notes, and the eight notes have become quarter notes. Wintory applies multiple imperfect augmentations throughout this piece in which the rhythmic patterns are expanded by similar but not identical amounts.
2) **Reharmonization**
   After a long journey through different tonalities, we are back at the melody being centred in the Bm chord. However, the melody is transposed to start on the D, the third note of the Bm chord, rather than the original B, the tonic. This simple change reframes the melodic movement into a major third rather than a minor third, as the melody now moves from a D to an F# instead from a B to a D, perhaps adding a touch of optimism as the journey is reaching its end. More importantly, on the two landing points of the ascending phrases, the melody is now supported by a G7 and a Cmaj7 chord for the first time. The restrain shown by Wintory to save this harmonic shift for this climactic moment makes the last bar of the melody especially powerful as the melodic centre of B now becomes the seventh of a Cmaj7 chord, shattering the B minor tonality and momentarily elevating the tonality by a semitone to C, perhaps mirroring the elevation of the player into their apotheosis.

Production tools – Remote recording

One of the many interesting aspects of this soundtrack is the lyrical and expressive performances of the musicians. The soundtrack was produced by combining a small ensemble of five highly skilled soloists that form the foundation of the score and the occasional use of larger string sections from The ***Macedonian Radio Symphonic Orchestra*** that were recorded remotely. The cello, that represents the player and is omnipresent, is played by the world-famous virtuoso Tina Guo whom Wintory also met in college.[7] There is also a solo flute (doubling as a bass flute), a solo harp, a solo viola and a solo French wind/brass hybrid instrument known as the Serpent. This is the only brass sounding instrument in the score and Wintory overdubbed it multiple times to create its characteristic lower rumble heard in the cave areas that are filled with magical dragon creatures.

It is interesting to compare the intimate production sound of the original *Journey* score with its epic reimagination by Wintory in a massive ensemble of 134 musicians including the London Symphony Orchestra and the London Voices choir in the game's 10$^{\text{-year}}$ celebratory album named **Traveller** (video example 44 with commentary by Austin).

Takeaway tasks

Task 1 – Arranging (moderate) – Create your own variation of the Journey theme

Take the melody from Figure 10.1 as your starting point and try to explore some of the techniques we discussed: augmentation/diminution of the rhythm of the theme, reharmonization using a different scale (both melodic/chords alterations), melodic inversions, creating free material with using fragments of the melody, different arrangements, different tempo, different meter, different transposition, different pitch registers. You might find it helpful to set a narrative context for yourself so you can use it as a guide. You can select an image from in game artwork/screenshots or you can set your own setting by using your imagination!

Notes

1 "First-Time Grammy Nominee: Austin Wintory."
2 Sua and Wintory, "Interview with Journey's Composer Austin Wintory."
3 Workman and Wintory, "Interview: Talking Journey's Majestic Music with Grammy Nominated Austin Wintory – Gamezone."
4 Hester and Wintory, "Why Austin Wintory Re-Recorded Journey's Soundtrack 10 Years Later."
5 Borkowski and Wintory, "Austin Wintory – Journey."
6 Napolitano and Wintory, "Exclusive: A Journey Through Journey's Soundtrack."
7 Oteiza and Guo, "Tina Guo – Interview."

Bibliography

Borkowski, Mariusz, and Austin Wintory. "Austin Wintory – Journey". *Gamemusic | Listen to Games*, 2013. https://gamemusic.net/austin-wintory-journey/.

"First-Time Grammy Nominee: Austin Wintory". *Grammy Awards*, 2014. https://www.grammy.com/grammys/news/first-time-grammy-nominee-austin-wintory.

Hester, Blake, and Austin Wintory. "Why Austin Wintory Re-Recorded Journey's Soundtrack 10 Years Later". *Game Informer*, 2022. https://www.gameinformer.com/2022/03/14/why-austin-wintory-re-recorded-journeys-soundtrack-10-years-later.

Napolitano, Jayson, and Austin Wintory. "Exclusive: A Journey through Journey's Soundtrack". *Destructoid*, 2012. https://www.destructoid.com/exclusive-a-journey-through-journeys-soundtrack/.

Oteiza, Gorka, and Tina Guo. "Tina Guo – Interview". *Soundtrackfest*, 2019. https://soundtrackfest.com/en/articles/tina-guo-interview/.

Sua, Michael, and Austin Wintory. "Interview with Journey's Composer Austin Wintory". *That Videogame Blog*. Accessed 1 October 2022. https://www.thatvideogameblog.com/interview-with-journeys-composer-austin-wintory/.

Workman, Robert, and Austin Wintory. "Interview: Talking Journey's Majestic Music with Grammy Nominated Austin Wintory – Gamezone". *Gamezone*, 2012. https://www.gamezone.com/originals/interview-talking-journey-s-majestic-music-with-grammy-nominated-austin-wintory/.

Chapter 11

The Last of Us (2013)

When less is more – Space and silence as storytelling devices

About the game – Story synopsis

To be able to follow this chapter it is necessary to give you a synopsis of the story in case you are not familiar with the game. Warning, major spoilers ahead! The game tells the story of Joel, a solitary and rugged smuggler who is trying to survive in a post-apocalyptic world after a Zombie pandemic takes over the USA. The game begins by showing us how Joel tragically fails to save his young daughter at the start of the crisis as she is unexpectedly killed during the riots. Then, we flash forward to a shattered world 20 years later and take control of Joel as he is trying to survive by completing different dangerous jobs. In one of these missions, he must transport a young teenage girl named Ellie across the desolate USA. During their hazardous journey, the emotional bond between them gradually grows to the point that he eventually sees her as his own daughter. When he discovers that her DNA is immune to the pandemic virus, he must decide between saving her life or sacrificing her in the hopes of developing a vaccine.

Fun trivia

Argentinian composer Gustavo Santaolalla follows a quite unconventional scoring process as he prefers to compose all the music based purely on the story and on conversations with the director, instead of relying much on the visuals. He has scored multiple films before they have even started shooting and the creative director of *The Last of Us* gave him total freedom to follow a similar approach with this game, which resulted in the music inspiring the designers to add things to the story that were not originally there.

> The way that I work is from the story, and having conversation with the director, not really from images. Even in the films I make most score before they even screen one frame. I don't consider myself as a film composer. 90% of composers work at the end, I find that very uncreative, I do my own take before the film or the game.[1]

DOI: 10.4324/9781003146872-12

How did the composer get the gig?

This was Gustavo Santaolalla's first game soundtrack, he was approached by the game designers based on his previous music and world-renowned film scoring background. Gustavo's solo album *Roncoro* in 1998 led to multiple prominent film directors such as Michael Mann and Alejandro Iñárritu asking to use extracts in their films, and eventually to a very successful career as a film composer with two academy awards for Best Original Score for *Brokeback Mountain* in 2005, and *Babel* in 2006. Despite his immense media scoring success, Gustavo does not consider himself to be a film or game composer and is an active touring musician.

Composition technique 1 – Use of space and silence as storytelling devices

"I like space and silence. I don't want to manipulate the audience, I 'd rather make it less obvious, I applied this to *The Last of Us*."[2]

Santaolalla's avoidance of using wall-to-wall music and minimal textures has been strongly evident in his prior film work but becomes even more pronounced in *The Last of Us* perhaps due to post-apocalyptic context of the story as well as the much longer duration of the game lasting approximately 15–25 hours for an average player. The exact amount of total music that is heard during that time is hard to calculate as it is dependent on how the game is played, but the official OST contains 56 mins of music. In my personal experience of playing the game, I found the use of music very sparse noticing multiple instances of over one hour of playtime with little or no musical accompaniment. Let us examine how this approach enhances the storytelling experience of the game.

Exploration

First, the use of silence allows the player to experience the sonic nuances of the desolate landscapes during the long journey across the USA. Without the use of music, it is easier to notice the emptiness of the environment and focus on ambient details such as the rain falling on tin rooftops of abandoned buildings, the wind rustling through grass and trees, the muddy footsteps and horse gallops, and of course the occasional zombie growls!

Minimal exploration music cues are triggered as the characters journey through key location points across the abandoned and desolate post-apocalyptic environments which provide us with insights on their emotional state. These cues are discrete and surprisingly short, often consisting of a few notes on a single instrument or a subtle atmospheric texture (video example 45, 30:20). The simplicity and scarcity of these cues is possibly reflective of the emptiness of this new world and the loneliness that the protagonists are experiencing. It is also possible that the absence of music is utilized to portray an absence of emotion from Joel. After the traumatic event of losing his daughter, he has become withdrawn, and basically just tries to keep on going as the years go by. However, as the journey continues and the bond between Joel and Ellie grows, there are also some musical hints of hope (ex: when they unexpectedly encounter a pack of Giraffes in video example 45, 9:07:00).

Second, it is precisely the lack of music that provides the necessary contrast needed so when rare musical moments do occur, they feel more meaningful and significant compared to the rest of the experience (video 45, 7:07:24). A lot of the exploration of the world is experienced without any musical accompaniment but there are pivotal moments that mark its development and those are accentuated by discrete use of music. This use of contrast is particularly useful for games in which players travel within large worlds and has been explored by game composers in completely different genres. As an example, the composers of the MMORPG *World of Warcraft* intentionally reserved the grand epic themes for special locations such as when you first stumble upon the main human city of *Stormwind* after a long session of silently wandering through a quiet forest.

Battle

A similar minimalistic approach is taken for the battle sequences which is quite rare for an action-based game. The soundscape of many battles often focuses on the use of diegetic sounds emerging from the battlefield such as terrifying screams (both human and otherwise), loud gunshots, and character Foley. This enhances the sense of realism and avoids the typical switch between battle and exploration music that can quickly become tiresome in other games. It also allows the player to focus on the directional sounds of the combat and be mindful of any vital information on the positions of the moving zombies (video example 45, 3:04:50). Sound plays a special role in the mechanics of the game as some Zombies use echolocation to detect the position of the player, meaning that if your movements are too noisy (ex: by running over some broken glass) you have a higher chance of being eaten alive! There is also a sense of awkwardness and tension that can come from silence itself, especially when you are trying to sneak around a highly threatening environment.

The use of music during battle sequences, when it does occur, is usually discrete. Many cues consist of drawn-out rhythms in a single percussive instrument, or a few understated drones (video example 45, 3:06:10). This minimal aesthetic here can also be interpreted as a reflection of Joel's withdrawn emotional state as he has become indifferent to the constant violence and sees it as a part of daily life. In more intense moments, the music reflects a growing sense of anxiety and tension, primarily through using dissonance, faster rhythmic pace, and bigger arrangements (video 45, 2:29:00). These higher intensity cues are reserved for the climactic action sequences which again creates a dynamic contrast with the rest of the game that is much quieter. However, even in the biggest fights the music never gets epic or heroic, but rather more intense, perhaps because Joel does not view himself as a hero (video 45, 3:07:45).

Composition technique 2 – Leitmotifs and storytelling

Aside from the use of silence and space, Santaolalla also incorporates leitmotifs to enhance other aspects of the storytelling. A leitmotif (from German meaning leading motif) is a theme or musical idea "whose purpose is to represent or symbolize a person, object, place, idea, state of mind, supernatural force or any other ingredient in a dramatic work."[3] Leitmotifs have been used for storytelling purposes from at least as early as the 19th century in operas, as well as in popular 20th century films (ex: the Force leitmotif in the *Star Wars* saga). There are many game examples in which leitmotifs are utilized to unify the story or suggest changes through variations of the musical elements, but *The Last of Us* is one of the most effective case studies, primarily because Santaolalla manages to tell the powerful story of the game with just two minimal leitmotifs. These two leading themes populate most of the central storytelling and cinematic moments in the game and are titled as *The Last of Us* and *All Gone* in the accompanying OST that can be found on Spotify. Tracks 3, 8, 13, 14, 18, 19, 20, 23, 27, 28 in the OST are different instrumental arrangements of just these two themes but it is noteworthy that the melodies themselves remain almost completely unchanged. Both themes are a very simple and minimal, yet their haunting melodies play a central role in portraying the emotional journey of the protagonists through the use of strong symbolisms. The simplicity of the musical language is a key factor as it makes it easy for the player to recognize them and follow any associations between the music and the narrative. As a contrary example, the two times Oscar winning score of the *Lord of the Rings* trilogy makes use of over 30 leitmotifs (one for the ring, one for Shire, one for the Wraiths, etc.) which allows the music to tell a more complex story, but it comes at a price of making it harder for the listener to identify what each motif represents and keep track of them as the story unfolds.

Analysis of the two leitmotifs

Have a listen to the first motif, *The Last of Us*, in video example 46 (0:20) which is the main theme of the game. As you can tell by looking at Figure 11.1, all the musical elements are extremely minimalistic. Motifs 1 and

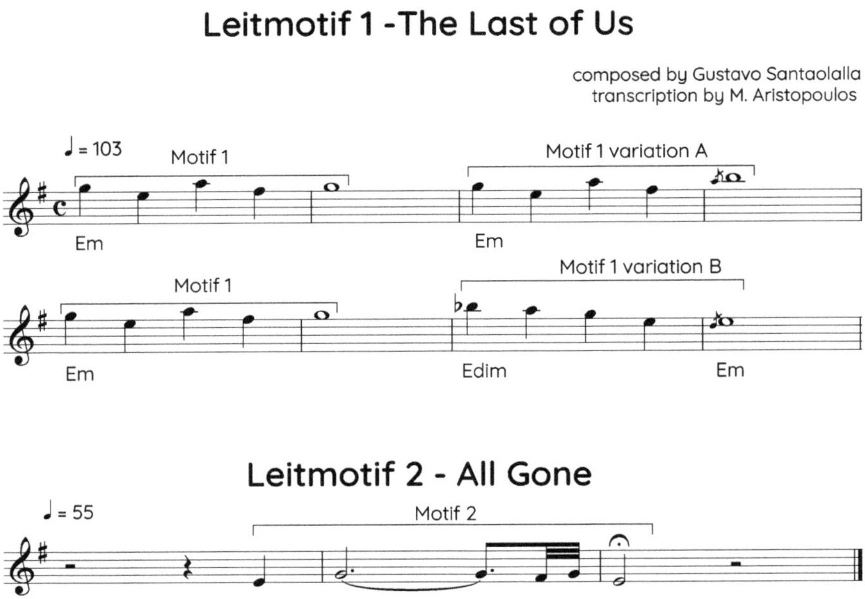

Figure 11.1 An analysis of melodic movement in the two leitmotifs.

variation A are identical except that one ends on the 3rd and the other one at the 5th of an Em chord. Motif 1 is repeated and answered by variation B which has an identical rhythm, but it temporarily replaces the Em with a diminished 5th before moving downwards to resolve back to Em. It is hard to imagine a simpler melodic structure than this as all the held notes are all part of the E minor triad. You can listen to the second leitmotif, *All Gone*, in video example 47. This is also based on just an E minor chord and is perhaps even simpler than the main theme. The phrase is repeated three times before it moves to a downwards arpeggio of Em.

How the two leitmotifs assist storytelling in the game

Now that you have familiarized yourself with the structure of these two leitmotifs, we can examine their use in the key narrative moments in the game in the analysis of Table 11.1. Video example 45 contains a walkthrough of the entire game with a length of approximately ten hours but you can use the indicated timecodes to locate each moment. The music underplays a lot of the horrible and dramatic events that occur, but on rare occasions it breaks the extended use of silence to highlight what is important: in a journey through a violent world full of grief and loss, these two protagonists only have each other, and the rest almost does not matter.

Table 11.1 Use of leitmotifs in key narrative moments of the game

Timecode	*Leitmotif*	*Symbolism*
00:00	*All Gone*	When the main menu screen is loaded, we hear the motif for the first time as a single melody in a lower octave. The music evokes a melancholic tone for the game before the story even begins.
02:30	*The Last of Us*	As Joel puts his daughter to sleep, we hear the motif for the first time as a single melody on a higher octave. This establishes a connection between the music and his relationship with his daughter.
16:00	*All Gone*	As Joel's daughter is shockingly and unexpectedly killed at very beginning of the story, we hear the theme on the violin with just one guitar chord as an accompaniment. The use of this theme during this remarkably sad moment creates a strong association between the music and the notion of loss. This connection will be repeated multiple times in the game.
17:00	*The Last of Us*	In a cinematic fashion the opening credits of the game come in surprisingly late, approximately 15 minutes into the story. We hear the motif in its biggest and fullest iteration. This establishes this music as the main theme and is one of the longest musical sequences in the game, the other one being the end titles.
57:00	No Music	Joel meets Ellie for the first time but notice that surprisingly, there is no music of any kind to support or foreshadow this fundamental narrative point. This makes dramatic sense, as Joel does not yet have any connection to Ellie and this encounter is just part of another meaningless job. What is particularly powerful here is that as the relationship and connection between them starts to grow from nothing, so does the use of music.
01:02:38	*The Last of Us*	When Ellie tells Joel that his watch is broken (it was gifted by his departed daughter), we hear the motif reminding us of the father-daughter connection. This is especially impactful as we had little music for almost over one hour of gameplay. It also serves as a time transition as Joel falls asleep, possibly dreaming of his lost daughter.
03:11:00	*All Gone*	As one of the people we encounter discovers the body of his partner, the motif comes back which is again used to signify loss during this personal moment.
6:54:43	*The Last of Us*	After a disagreement, Ellie goes missing and Joel rides to look for her while we hear this motif in a full arrangement. Until that moment, both characters tried to appear rather indifferent towards each keeping an image of toughness and grit. However, as Joel rushes into the forest to search for Ellie, the same music that has so far represented his daughter is heard, but for the first time in reference to Ellie. This is a powerful moment which discretely implies that Joel now sees Ellie as his own child, as this has now become her theme too.
9:48:28	*All Gone*	We hear the motif to symbolize the possibility of loss As Joel is trying to save Ellie's life, we hear the motif associated with loss that subtly symbolizes the possibility of her not making it through.
9:53:00–end	*Home & The Path*	During the ending of the game, we get neither of the two leitmotifs, but instead two new themes, perhaps to suggest a more hopeful tone for the new chapter that begins in Joel and Ellie's lives.

Production tools – Guitar based techniques

To accompany this post-apocalyptic story Gustavo relied on unconventional uses of acoustic and traditional instruments from South America, which has been a characteristic of his sound as an artist and as a film composer. Two of the dominant instruments used are the ronroco and the charango (Figure 11.2), two small guitar like instruments from the Andes which have a long musical tradition dating back to at least the early 18th century. They have five double strings and were traditionally constructed from wood at the top and an armadillo shell for the back, but this is no longer the norm. The ronroco has a lower pitch and a longer sustain than the charango, and Gustavo plays it in an unconventional using finger picking to play a melody and accompaniment at the same time. It has been a pivotal instrument in Gustavo's film music career that thrived following the release of his first album based entirely on ronroco compositions.

Additionally, Gustavo used an electric resonator dobro guitar with all strings tuned down two whole notes from a EADGBE tuning to a

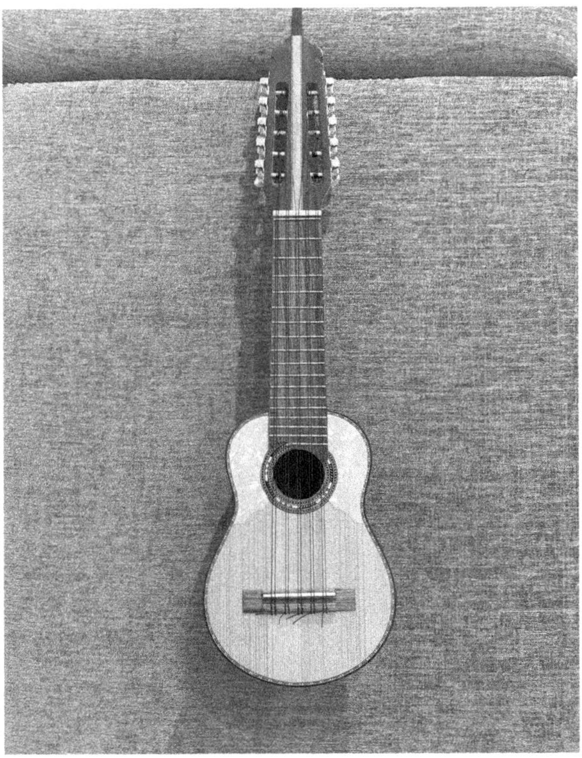

Figure 11.2 A photo of a ten-string Charango guitar from the Andes. The photo was provided by the film composer Americo Martin.

CFBbEbGC. By using such an extreme tuning the loose strings create a darker, almost out of tune timbre. This was particularly useful, as he also used a violin bow on the guitar strings themselves, an unusual technique which produces a very interesting harmonic timbre that sounds like a harsher version of a cello, as the strings themselves are metallic (stainless steel). This technique has also been explored by film composers such as Taylor Bates in the *300* soundtrack and has even led to the birth of a new instrument, the guitar viol.

Finally, the orchestral ensemble recorded for the game was also modified to use predominately lower pitch instruments to fit Santaolalla's vision for the game. The string section had no violins and used only violas, cellos, and double basses, while the wind section used predominately bass clarinets and bass saxophones. In terms of percussion, many of the rhythmic textures are generated from uses of scrap items such as PVC pipes, springs, used cans, custom bells, and buckets leading to a very different production aesthetic for the action cues compared to the clean drum ensembles found in Hollywood blockbusters and AAA games.

Takeaway tasks

The two tasks can be combined or completed separately.

Task 1 – Composition (moderate) – Write a minimalist leitmotif

You can write a short theme to represent any idea, person, or place you find inspiring from the game. To fit the musical language of the rest of the soundtrack, try to make use of the minimal aesthetic of the other leitmotifs by using minimal instrumentation, harmony, textures, structure, and development. Remember that sometimes less is more!

Task 2 – Composition/production (challenging) – Write a theme that uses unconventional guitar techniques

Create a short composition that relies on unconventional uses of guitar techniques such as the ones discussed in this chapter. You can start by experimenting with using a bow on the guitar strings. You need to use string rosin on the bow to get a proper sound. What will also help is if you add reverb and/or delay in order to enhance the sustain, as well as tuning down your strings so they have more slack. You can try this on an acoustic or an electric guitar, or bass. If you do not have access to a violin bow, do not worry, you can try a drumstick or even a pencil! Just watch video example 48 for inspiration on similar experimental techniques.

Notes

1 Reese and Santaolalla, "Gustavo Santaolalla and the Last of Us."
2 Reese and Santaolalla, "Gustavo Santaolalla and the Last of Us."
3 https://www.oxfordmusiconline.com/grovemusic/view/10.1093/gmo/9781561592630.001.0001/omo-9781561592630-e-0000016360.

Bibliography

Reese, Emily, and Gustavo Santaolalla. "Gustavo Santaolalla and the Last of Us". Podcast. *Top Score*, 2019. https://itunes.apple.com/us/podcast/top-score/id434473316.

Chapter 12

Alien Isolation (2014)

In space none can hear you scream – Controlling tension with a vertical layers system

About the game

A survival horror game inspired by the original film by Ridley Scott. Set a few years after the events of the first film, the game follows Amanda Ripley, daughter of Ellen Ripley, as she tries to understand the disappearance of her mother whilst stranded on a large space station. It takes the form of a first-person stealth horror with linear, closed world level design.

Fun facts

When the game is played on a console it can track your real-world noise levels using the Microsoft Kinect and PlayStation camera hardware. According to the audio designing team: "if you scream on your sofa you'll give away your position in-game!".[1]

How did the composer get the gig?

The music for the game was written by *The Flight*, an East London based composing duo consisting of Joe Henson and Alexis Smith who started collaborating in 2005. The duo already had some media credits primarily in UK TV series and documentaries before moving into writing for games. They were approached by *Creative Assembly*, one of the biggest UK game development studios, and asked to pitch for a demo of the game.[2] After the success of *Alien* they have contributed additional music to many well-known AAA titles such as *Assassin's Creed Odyssey* and *Horizon Zero Dawn*.

Composition technique – Controlling tension with a vertical layers system

As you would expect, a survival horror scenario in which you are locked up in a claustrophobic spaceship while being hunted by a highly evolved predatory alien will inevitably lead to some grippingly anxious moments. In the original *Alien* film of 1979, the composer could carefully plan how

DOI: 10.4324/9781003146872-13

to create maximum impact precisely at the right moments as the dramatic development and timing of events were predetermined. However, when the same scenario is realized within a video game format it will unavoidably require a radically different musical approach as the fluctuating levels of tension are controlled by the relatively unpredictable actions of the player. One common interactive technique that can be remarkably useful in this context is the use of a musical system based on vertical layers.

Interactive composition 101 – Vertical layers

Vertical Layers is a simple yet powerful interactive technique that is based on breaking down a musical arrangement in several vertical layers (also known as stems) which can be added or removed from the arrangement in response to gameplay changes. These changes can be tracked in real-time using a range of numerical values such as "number of enemies present" as well as specific triggers such as "health is below 10%" or "final boss has arrived".

In theory, any composition can easily be fragmented to smaller parts which can be used as layers if it is recorded in a multi-track format. However, for a vertical layer system to work successfully it needs to be designed and tested accordingly so it can be function well in a range of scenarios. For example, a single layer might play on its own for a long period of time during exploration mode or a peculiar layer combination might suddenly be triggered. Therefore, it might be more effective to think of layers as interdependent compositions that can work vertically with each other in musically interesting ways, rather than single instrumental tracks that are extracted from a bigger arrangement.

The composers of *Alien* wrote the music with a vertical system in mind:

> We worked very closely with the developer on the music system. This was based on many factors – the environment, the state of play, the proximity of the Alien, its 'state' – whether it knows you are there, is facing you etc. They had an idea of what they wanted the music to do in this game, and we had to find a way to make it happen.[3]

After the music was composed and delivered to the audio team in a "kit" format, the audio files were implemented in a sophisticated layer system designed in Wwise using RTPC (Real Time Parameter Controls) to control the volume of individual layers.

Let us examine some of the behaviours of the vertical layers in the game. In video example 49, (4:09:30–4:10:41) you can observe a gameplay capture of the alien searching for you within a confined space. As the creature moves around you the system reads numerous parameters from the game in real time (obstructed distance from the *Alien*, player stealth, total threat) and accordingly adds or removes instrumental layers from the mix to build up or reduce the musical tension.[4] Notice how the lower pitched layer remain consistent to provide continuity but as the alien comes closer you the first time a chaotic string texture is added and then removed as he walks away. The second time he approaches the chaotic strings return but are enhanced by more strings in a higher register along with another layers of distorted synths (Figure 12.1).

This technique is undoubtedly effective in synchronizing the audio-visual changes of suspense but if the layers were always directly connected to the same parameters (ex: the distance between the alien and the player) then the musical build-up would quickly become predictable. Moreover, after you will feel familiar enough with the music to become aware with this convention, it would become detrimental to the playing experience as the music would always give away the *Alien*'s position, thus ruining any elements of surprise. To counteract this problem, the audio team designed a range of different layer behaviours that the music engine can pick to avoid making the musical mechanisms easy to perceive and to constantly keep you on edge. Something I observed while playing the game is that the music is also

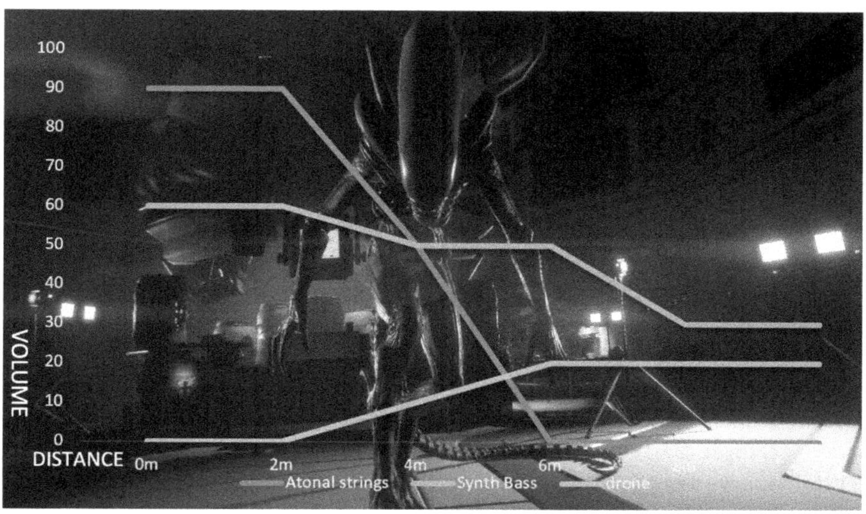

Figure 12.1 A chart illustrating an example of how the audio volumes of three musical layers respond as the distance between the player and the alien is altered.

being used to build a false sense of security that can be turned against you right when you begin to trust it. In video example 49, 6:18:12, you can notice that there is a lack of scary music, the arrangement is calm, and everything feels temporarily safe, but as soon as you turn around you suddenly notice a familiar face staring at you!

The vertical layers do not only adapt to the actions of the alien; some of the action music adapts to the AI state of hostile humans as well as the amount of danger of various tasks that you are pursuing according to three levels of intensity.[5] In video example 49, you can observe how a layer of very heavy percussion is added to the rest of the mix the moment the player activates a particular objective (06:25:00), which is then subdued once that objective is completed (06:25:41). Some of these jump scares are also synchronized using gameplay triggers and a closely related interactive technique that is known as music stingers. The main difference is that stingers are usually much shorter, and do not necessarily have to be synchronous to the rest of the layers as they can be triggered at any point. (For more information on stingers see Chapter 18: *Tomb Raider*).

Production tools – Extended orchestral techniques

The game's soundtrack includes over three hours of music that was inspired from three key themes from Jerry Goldsmith's score from the original film *Alien*.[6] During the recording of the game's soundtrack, the composers worked with the Chamber Orchestra of London at Air Studios which included some of the original players from the iconic film score recording of 1979! This opportunity granted the composers access to some inside information on the original techniques used by Goldsmith that they would not have been able to work out without their input. They recount:

> One example was the 'Alien Whale' sound; we had been trying various techniques to replicate this, from bowed drums to rubbing superballs on the underside of a piano. One of the original players told us they thought it was a conch shell. We used all three of the techniques throughout the game![7]

You can listen to the *Alien Whale* technique in the main menu music in video example 50. You might also notice that the strings sound rather different than your usual orchestral score. This sound was achieved by using several different string techniques borrowed from 20th century orchestral music that are usually referred to as "extended" to differentiate them from traditional instrumental writing. Goldsmith's score, that was allegedly not used as he wanted in the film, was filled with interesting uses of extended orchestral techniques ranging from screeching atonal strings to using multiple household objects on a prepared piano.[8] In Table 12.1 you can observe some of the extended string techniques from video example 50 that are also utilized often throughout the soundtrack.

Table 12.1 Examples of extended string techniques used in the *Alien* soundtrack

Technique	*Description*	*Notation*
Sul pont	Bowing close to the bridge. Produces a thinner less rich tone.	Written Instruction above the stave
Bow on bridge	Not the same as Sul Pont which is near the bridge. On the bridge produces a high screeching sound with little discernible tone.	Written Instruction above the stave
Sul Tasto	Making bow contact with the string close to the fingerboard. Produces a soft, almost muted tone.	Written Instruction above the stave
Con Lengo	Usually played with a bouncing staccato motion by directing the player to make contact with the strings with the wood in the back of the bow (ex: the triplet motif in the opening of Mars from Holst's Planet's).	Written Instruction above the stave
Bartók Pizz	Often called "snap pizz", this pizzicato technique requires players to get their finger of underneath the string and pull it upwards, releasing the string so that the string snaps against the fingerboard.	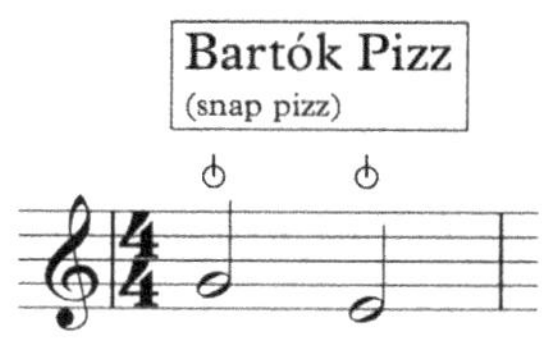
Harmonic Glissando	Used most notably by Stravinsky, a harmonic glissando runs through the harmonic series on any given string and is performed by the player running their finger up and down the string. It can be combined with an irregular tremolo and more erratic left-hand movement to create high, screeching sounds.	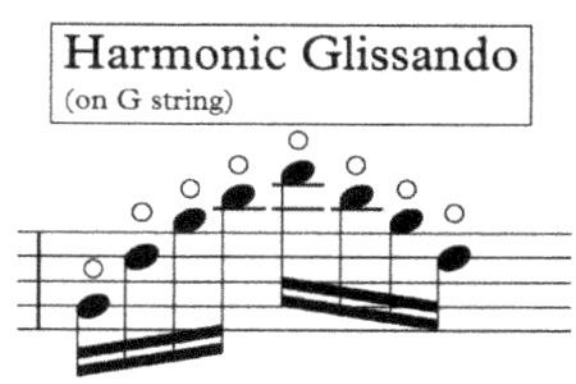
Sempre Sul Pont Glissando (Screeching Strings)	Similar to the technique above without focussing on the harmonics and instead focusing on fast or irregular tremolo played near the bridge. Heard most notably in George Crumb's "Black Angels".	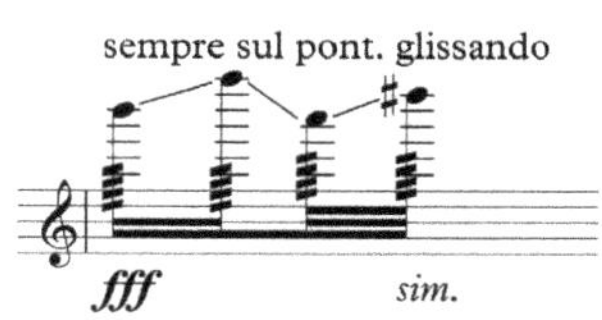

Takeaway tasks

The following two tasks can be optionally combined.

Task 1 – Composition (challenging) – Write a string-based composition that explores some of the extended techniques discussed in this chapter

Ideally you want to do this task with an instrumentalist as these techniques are rarely found in your typical sample libraries unless they are on the higher end of the professional range. However, you can find some of them in specialized libraries such as Spitfire Labs Strings 2 that is available for free on the company's website! An alternative solution is to do this task by using a prepared piano in which you place various objects in between the strings, similarly to Jerry Goldsmiths creative use of a rubber ball to produce the *Alien Whale* sound, but be careful not to damage the instrument.

Task 2 – Implementation (moderate) – Create a vertical layer arrangement that will react to four different levels of tension

Here are some tips to remember: a single layer could contain multiple instruments, all possible combinations should be musically interesting, layer 1 should work on its own and each additional layer should clearly contribute to a rising level of tension. If you are not familiar with an implementation software in which you could test your system (ex: Wwise), you can easily simulate this in your DAW by fading in and out layers at a random timing to emulate hypothetical game triggers.

Notes

1 Bullock et al., "Listen or Die."
2 Larson, "Gaming Music: ALIEN: ISOLATION Interview with the Flight (Joe Henson & Alexis Smith)."
3 Larson, "Gaming Music: ALIEN: ISOLATION Interview with the Flight (Joe Henson & Alexis Smith)."
4 Bullock et al., "Listen or Die."
5 Bullock et al., "Listen or Die."
6 Usher, "Alien Isolation Interview: How Composers Evolved a Legacy."
7 Larson, "Gaming Music: ALIEN: ISOLATION Interview with the Flight (Joe Henson & Alexis Smith)."
8 "Alien – The Complete Original Score."

Bibliography

"Alien – The Complete Original Score". *Jerrygoldsmithonline.Com*, 2022. http://www.jerrygoldsmithonline.com/alien_review.htm.

Bullock, Byron, John Broomhall, James Magee, and Haydn Payne. "Listen or Die". *Audiotechnology*, 2015. https://www.audiotechnology.audiotechnology.com/features/listen-or-die.

Henson, Joe, and Alexis Smith. "The Music of Alien: Isolation – Interview with the Flight". *MCV/DEVELOP*, 2015. https://www.mcvuk.com/development-news/the-music-of-alien-isolation/.

Larson, Randall. "Gaming Music: ALIEN: ISOLATION Interview with the Flight (Joe Henson & Alexis Smith)". *Buysoundtrax.Com*, 2014. http://www.buysoundtrax.com/larsons_soundtrax_12_2_14.html.

Usher, Will. "Alien Isolation Interview: How Composers Evolved a Legacy". *CINEMABLEND*, 2014. https://www.cinemablend.com/games/Alien-Isolation-Interview-How-Composers-Evolved-Legacy-68091.html.

Chapter 13

Mario Kart 8 (2014)

Music as an information device

About the game

Mario Kart 8 is the most recent instalment of Nintendo's kart racing franchise originating from *Super Mario Kart* for SNES in 1992. Players choose one of the iconic characters from the Mario universe and race in colourful racetracks filled with magical objects that can be used to boost their vehicle or malevolently against other players to hinder their progress. It is allegedly responsible for many ruined friendships!

Fun facts

Mario Kart 8 is the best-selling Wii U game of all times with more than 8 million copies sold, and the best-selling Switch game of all times with more than 47 million copies sold.[1] Even though it is available exclusively for these two Nintendo consoles, it is still the best-selling racing game of all times in any platform, and the no.7 best-selling game overall (with Minecraft being number 1).

How did the composers get the gig?

Mario Kart 8 was developed by a large team of composers over many years. All the composers have had a long working relationship as full-time in-house employees of Nintendo Japan: Kenta Nagata joined in 1996, Ryo Nagamatsu in 2006, Shiho Fujii in 2007, Atsuko Asahi in 2010, and Yasuaki Iwata in 2013. Many of these composers have spent their entire career working exclusively for Nintendo and are also responsible for other beloved musical gems such as the *Zelda* series, *Super Mario*, and *Animal Crossing*. This business model of relying on in-house composers is relatively uncommon in the gaming industry outside Japan, as most studios rely on freelancers hired on a project-by-project basis.

DOI: 10.4324/9781003146872-14

Composition technique 1 – Music as an information device

The music of *Mario Kart 8* interacts with the gameplay in multiple creative and entertaining ways that are not just decorative but provide important feedback on what is happening in a race. In fact, someone familiar enough with the soundtrack can tell quite a lot about what is happening in a race even with their eyes closed! Having these additional layers of musical feedback on top of the fast-paced visual information can make the game easier to play and quicker to learn, especially for younger audiences. The music possibly plays such an integral role in the game that Nintendo does not give you the option to play it with the music turned off. Let us examine some of the different bits of information that the music communicates.

Racetrack location

Mario Kart 8 is the first game in the series to have each of the 32 racetracks feature its own exclusive level theme, making it clear to the player where the race is taking place. Some of these themes have been developed from previous games in the franchise and others are unique.

Race stages

The music clearly emphasizes each stage of the race: there is a cue for previewing the racetrack, another short motif for preparing players that the countdown is about to begin, the main level theme for the actual race, and a cue for the outcome of the race. Depending on the position you finished in the race you get one of three different musical themes, one for 1st place finish, one for 2nd–6th, and one for 7th–12th place.[2]

Specific zones within racetracks

Beyond being able to recognize levels by their musical theme, there are also specific sublocations within many racetracks that can be identified by their imaginative music embellishments when the player drives through them. In video example 51 you can see the 3DS Music Park racetrack which contains various road surfaces made from keyboard instruments (ex: piano, xylophone, glockenspiel). As you drive through each surface chromatic runs from that instrument are added to the arrangement that are perfectly synchronized with the tempo and pitch of the level music theme. Another great example that contains one of the game's most popular music themes can be seen in video example 52 that showcases a drive through the *Dolphin Shoals* racetrack. As soon as the player leaps out of the underwater section (0:53) the melody is replaced by a blistering live saxophone solo that falls perfectly in time, and the music modulates upwards. To make the switch sound natural, there is a splashing sound effect that masks the crossfade.[3]

Movement of in-game objects

Many in-game objects have their animations synced to the music and therefore give away the timing of their movement. For example, the giant enemies shaped as musical notes on the Melody Motorway level are jumping in time with the music; or the piranha plants on Piranha Plant Pipeway snap their teeth on beats 2 and 4!

Position in the race

The game uses an innovative system called **front running beats** to audibly indicate that a player is in first place. The 'beats' are a rhythmic top layer, usually consisting of fast punchy kick, with off-beat hi-hats that trigger when a player has taken a big enough lead in the 1st place from the other karts. This increases the pressure on the leader whilst alerting other players in multiplayer mode that the leader is getting away. This drum layer is stripped if you are hit by a shell or when you fall from the course. In video example 53, 1:27, you can observe how these front running beats are added as the player's lead increases and in video example 54, 0:10, how they are removed as the player suddenly crashes and loses the lead.

Final lap

When you reach the final lap of the race the level music becomes more frantic by speeding up the tempo by 30% and pitch shifting upwards by one semitone (video example 52 from 2:02).

Gameplay events

With a few exceptions such as the car sounds and the voices, most SFX in the game have an innately musical quality to their design (ex: Star Item received is a quick Dmaj7 – Cmaj7 chord pattern). These musical SFX are usually added on top of the level music and can communicate various pieces of information to the player in an entertaining way that aids the synergy between SFX and music and avoids overwhelming the player with additional text in busy races. The practise of musical SFX design is typical of Nintendo (also see Chapter 3 on *Zelda*).

Gameplay states

There are some gameplay events that are reflected with the addition of production effects in the music. For example, when a player deploys a lightning strike, the audio echoes this by applying a fast-moving flanger effect to the top melody lines of the music, as well as the engine notes and character noises. The further you are in front, the longer these audio changes seem to last, indicating to other players that the leader is still slow, and therefore easier to catch up.

Composition technique 2 – A masterclass of key modulation

Changing the tonal centre of your theme through modulation can be a useful way of adding further interest to your music. *Mario Kart 8* offers a true masterclass on the use modulation as the music constantly changes keys, often multiple times within very short thematic segments. Perhaps the music makes such extensive use of modulation to reflect the frenzied pace of kart racing. Some of the uses of modulation are part of the written music, while others occur as responses to gameplay events (ex: pitch shifting by one semitone on the last lap). There are many techniques used to modulate, let us examine some of the most common ones found in the game.

Modulation 101 – Circle of Fifths

You can use the Circle of Fifths (Figure 13.1) as a tool to find the signature of different keys and to understand the relationships between them when planning your modulations. Generally, the closer two keys are in the circle, the more notes they will have in common. If you take any key and look at its neighbouring keys on the left and right, you will notice that their key signatures only have one note difference (one sharp moving clockwise and one flat moving anti-clockwise). You can also see the minor keys that share an identical key signature with their relative majors.

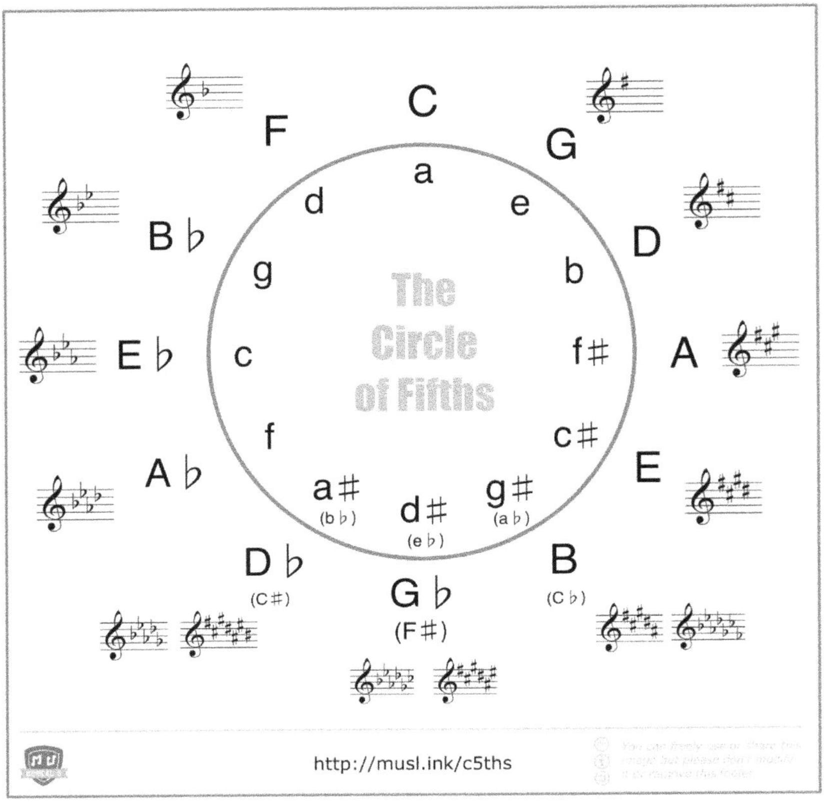

Figure 13.1 A diagram of the Circle of Fifths showing the relationship between the 12 major keys, their associated minor keys, and their key signatures.[4]

Pivot chords (common chords)

One easy way of using the circle of firths to change keys smoothly that is used in classical music is by using pivot chords. These are chords that are common between both keys and can be used to link them. The closer the keys are to each other in the circle the more chords they will have in common that could be used as a pivot. For example, from the key of C major to G major, there is only one note difference (F#) which results in four of the chords being identical (I, iii, V, vi). However, if the modulation is towards a distant key, for example from C to Db, there are no common chords between them that can be used as a pivot.

Secondary dominant

Although common chords are used frequently in *Mario Kart 8*, this technique is thrown out of the window in the main theme! Listen to the first ten seconds of the theme in video example 55. The theme starts in G major (a key signature of one sharp) and only five seconds later it has already modulated to Ab major, a completely unrelated key with a key signature of four flats that has no common chords. Yet, this big change stills manages to sound smooth as these unrelated keys are connected by a dominant chord that comes in 00:05" in the brass right after the bass solo. This connecting dominant chord (Eb7) clearly does not belong to the first key, but it functions as a dominant seventh chord in the context of the new key of Ab and prepares its arrival. This modulation technique is known as a **secondary dominant.**

Phrase modulation

Another modulation technique that is used constantly in this soundtrack is phrase modulation. Repeating a phrase directly in a new key can be a great way of bridging big tonal jumps as the linear harmonic relationship between the notes of the phrase is already familiar to the listener, and the new key can provide a fresh context. This idea is frequently utilized in in pop music, when you have a repetition of an already familiar chorus modulating higher towards the end of a song. Listen to section B of video example 55 from 0:33 to 1:04. The music moves between multiple modulations by using repetitions of the same phrase but each time in a new key. The same technique can be found in *Dolphin Shoals* where the music uses a tonal sequence to modulate upwards by one semitone from Fm7 Bb7sus4 to F#m7 B7sus 4 (video example 56, 0:15–0:25), and also in the *Mount Wario* theme (video example 57, 0:30–01:00) where the music constantly transposes upwards through phrase modulations.

Parallel modes

Finally, another technique used is parallel modes (for more info on how to build modes see *Mortal Kombat* Chapter 7). Although this might not technically be a modulation, as the tonal centre remains the same, the use of parallel modes is frequently used in jazz solos to quickly move between different scales that are compatible with the underlying chord notes and thus adding further harmonic interest. For example, if you have a C7 chord (CEGB) your melodic solo can be in C major scale, but it can also move to C Lydian as the F# does not clash with the harmony. A good example in Mario Kart are the guitar solo in the first 30 seconds of the main theme which temporarily moves to Ab Myxolidian by flattening the 7th note of the major scale.

Production tools – The Mario Kart Band

Another feature that makes this soundtrack so much fun to listen to is that most of the music was recorded with a live band, the Mario Kart Band! This is an all-star group consisting of some of the top Japanese jazz musicians, many of them coming from the band *Dimension*.[5] As you can see in video example 58, the performances are captured live, but the musicians are separated by acoustic panels to allow the isolation of individual stems that can function interactively in the game. It is worth mentioning that there is a plethora of instruments and specialist musicians added for recording each racetrack that go far beyond the usual jazz big band instrumentation. Some examples from different racetracks include *Twisted Mansion* that features an organ and a theremin, *Sweet Sweet Canyon* that uses an accordion and a recorder among many peculiar percussive instruments, *Dragon Driftway* that features a Chinese Erhu, and *Thwomp Ruins* that features an Indonesian Gamelan!

Takeaway tasks

Task 1 – Analysis (easy) – Analyse the use of music as a source of information in another game of your choice

What is the music telling you in direct or indirect ways? For example, you might notice that in an action game the music clearly indicates when you have been detected by enemies. Make a short list of your observations with references to gameplay conditions. Some of these might be obvious (ex: win/lose music) but there might be a lot of more subtle information that is given indirectly (ex: a new instrument being introduced when your health is low, etc.).

Task 2 – Composition (challenging) – Write a theme for a Mario Kart 8 *level of your choice that makes significant use of modulation*

You can look at the circle of fifths in Figure 13.1 to help you plan your modulations. Remember that you can use pivot chords, phrase modulation, and even parallel modes. If you have access to live musicians (or even better a live band), you can use them to co-compose and record your theme. You might find it useful to provide the musicians with an overall chord sequence and main melody but allow them the freedom to contribute their own improvizations as they like that you might then incorporate in the final mix. Remember to isolate each sound source as much as possible so you can have more options during the mixing and (potential) implementation stages.

Notes

1 "Sales Data – Top Selling Title Sales Units."
2 "List Of Mario Kart 8 Media."
3 Schilling and Nagata, "The Music of Mario Kart 8."
4 "The Ultimate Guide to the Circle of Fifths."
5 "Liner Notes – Mario Kart 8 Original Soundtrack."

Bibliography

"Liner Notes – Mario Kart 8 Original Soundtrack". *Super Mario Wiki*. Accessed 1 October2022.https://www.mariowiki.com/Mario_Kart_8_Original_Soundtrack.

"List of Mario Kart 8 Media". *Super Mario Wiki*. Accessed 1 October 2022. https://www.mariowiki.com/List_of_Mario_Kart_8_media.

"Sales Data – Top Selling Title Sales Units". *Nintendo Co., Ltd.*, 2022. https://www.nintendo.co.jp/ir/en/finance/software/index.html.

Schilling, Chris, and Kenta Nagata. "The Music of Mario Kart 8". *Kotaku Australia*, 2014. https://www.kotaku.com.au/2014/06/the-music-of-mario-kart-8/.

"The Ultimate Guide to the Circle of Fifths". https://www.musical-u.com/. Accessed 17 October 2022. http://musl.ink/c5ths.

Chapter 14

Apotheon (2016)

Recombinant cells – A generative technique for producing musical variation

About the game

A 2D platform indie game inspired by ancient Greek art and mythology (Figure 14.1). The game was developed by Alientrap, a small independent Canadian studio and was selected by Sony as a PlayStation Plus release. Both the game and the soundtrack have been a commercial success with multiple awards and high sales for an indie game. The music was nominated for best soundtrack of 2016 year by the Canadian Game Awards and was surprisingly even placed an honourable mention of the best PlayStation Soundtracks of the Decade by *Push Square* magazine![1]

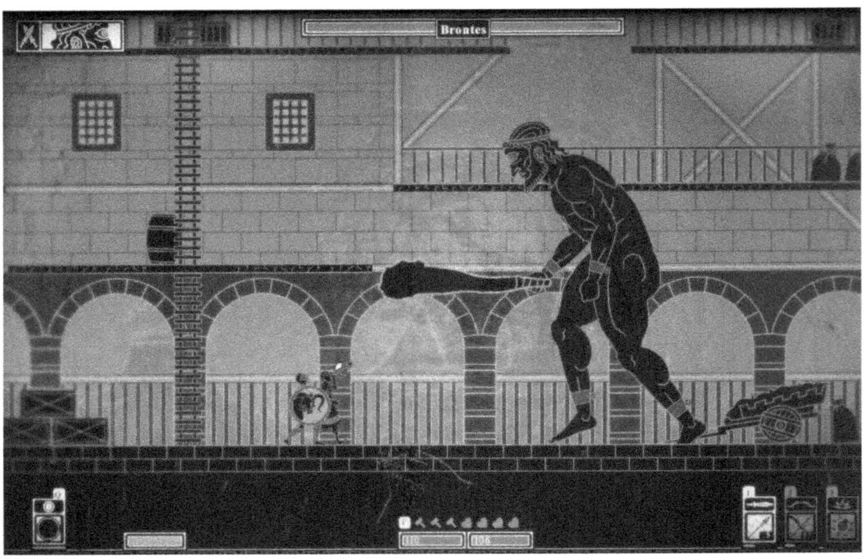

Figure 14.1 The art style of the game is based on the black-figure pottery paintings from ancient Greece.

DOI: 10.4324/9781003146872-15

Fun trivia

This soundtrack was composed by me! It was an incredible honour to have extracts of the music performed to accompany the Olympic Flame Initiation Ceremony in Athens for the 2019 Special Olympics (video example 59).

How did the composer get the gig?

Prior to *Apotheon* I had written ancient Greek music for various TV documentaries in Greece, and I was quite active at writing music for mod games I found online. One of the *Total War* mods I wrote the music for was called *Troy* and I had placed its entire soundtrack on Indie Db as a free download for non-commercial uses. The developers of *Apotheon*, unbeknownst to me had been using this music as a temp track during their early stages of prototyping *Apotheon* and reached out to me to request a licencing deal. I happily accepted the offer but also proposed to write new music to their specifications which led to a successful collaboration that lasted almost three years.

Composition technique 1 – Recombinant cells

Recombinant cells is a relatively simple generative technique that can be used for producing musical variation in game soundtracks. It works by recombining musical motifs in real time using a range of probabilities that are connected to gameplay parameters. I developed this technique in collaboration with *Apotheon*'s programmer Lee Vermeulen to avoid playing identical loops of the same music in key areas of the game in which players might spend large amounts of time. This idea was inspired by the aleatoric techniques of Mozart's musical dice games from the 18th century that used the roll of two dice to generate waltz compositions from two sets of pre-composed motifs that could produce 759,499,667,166,482 unique yet similar waltzes.

A similar recombinant approach has been explored by other composers in earlier games such as *Ballblazer*, *Times of Lore*, *Legend of Zelda: Ocarina of Time*, and *Red Dead Redemption*. One primary difference with the use of this technique in *Apotheon* is that the recombination of musical phrases can occur simultaneously across multiple layers that are affected by gameplay parameters and conditions. The exact rules of the generative system in the game were adapted to the design of each level, but each of the musical phrases needed to be composed specifically with those rules in mind to generate a musically interesting output.

How recombinant cells work in the opening level

The most elaborate version of this technique can be found in the opening level of the game called *The Village of Dion*, a large open area that included many sublocations that players could visit and complete quests in almost any order. A walkthrough demonstrating the entire level is provided in video example 60. The recombinant music system consists of a collection of 37 different musical phrases (the cells) that were composed to correspond to all the possible gameplay developments for this area. Each cell has its own default occurrence probability, trigger conditions, and layer group according to their function in the arrangement (ex: melody). For example, as you can see in Table 14.1. The qanun, a type of traditional string instrument found in Greece, is part of one layer that contains five phrases with an equal 10% change of being triggered, and there is a 50% chance that no qanun phrase will be played thus reducing the number of instruments being played in the arrangement. Similarly, the battle string layer contains five phrases with an equal 20% chance of being triggered but only if there is a battle occurring in the game. The length of most cells was usually proportional to each other to ensure they will remain in sync, unless some level of rhythmic anarchy was desired, an idea that was used extensively in another level (The

Table 14.1 The default occurrence probabilities of the cells in two mini quests from *The Village* area

FIND THE APOTHECARY (ACADEMY) QUEST		
Cell Name	*Cell Probability*	*Layer Probability*
ambience-01, ambience-02	50%	100%
qanun-f1, qanun-f2, qanun-f3, qanun-f4, qanun-f5	10%	50%
percussion-01, percussion-02, percussion-03, percussion-04, percussion-05	20%	100%
battlestrings-01, battlestrings-02, battlestrings-03, battlestrings-04, battlestrings-05	20%	TRIGGERED when there is battle
FIND THE BLACKSMITH (BLACKSMITH'S HOUSE) QUEST		
ambience-01, ambience-02	50%	100%
anvils-01, anvils-02, anvils-03	33.3%	100%
battlepercussion-01, battlepercussion-02, battlepercussion-03, battlepercussion-04, battlepercussion-05	20%	100%
battlebrass-01, battlebrass-02, battlebrass-03, battlebrass-04, battlebrass-5	20%	TRIGGERED when there is battle
Timpani-f1	100%	TRIGGERED when there is battle

underground caves in the Forest). The harmonic language of the cells was heavily based around different parallel modes to make it easier to match overlapping cells together as they would share the same tonal centre (for more info on parallel modes see Chapter 7 *Mortal Kombat*).

What gives this system more value than a curated random music generator is that all the cell probabilities are not permanently fixed but are dynamically altered according to what is happening in the game (see Figure 14.2). As the player completes various objectives and moves through the map, different cell probabilities are modified over time resulting in a much more adaptive soundtrack that transitions smoothly between gameplay changes. For example, if the player moves to a new location in the map, a set of melodic cells will replace the previous melody, but the rest of the layers might overlap between the two sets creating a hybrid arrangement. Likewise, if the final boss (*The Tyrant*) quest for the area is triggered, all the selection probabilities of the cells will be altered to favour cells that evoke an epic mood, and any cells that are not relevant will be removed from the selection pool by having their probability set to 0%. In this way, the music generation is closely driven by the gameplay action, but it (almost) never repeats an identical arrangement. After eight years of having created this generative/interactive theme, although its execution is far from perfect, I still get occasionally surprised by interesting new re-combinations that have morphed the original material in a familiar, yet ever-changing shape that prolongs the life of the music and the re-playability of the game. If you are interested in learning more about this technique you can read my PhD thesis that is available for free in The British Library Thesis database.[2]

Production tools – A virtual choir singing in ancient Greek

In the grand finale of the game, you must face the king of the Greek gods, Zeus, to reach your apotheosis (video example 61). During this battle I wanted to have the biggest choir sound imaginable. I was inspired by the Viking choir in *Skyrim* that was achieved by overlaying multiple takes of the same performance but due to budget constraints I used 12 different layered instances (as many as the gods of Olympus) of *East West's Symphonic Choirs* virtual choir. The MIDI data was triggered through the *WordBuilder* plugin which allows users to type in syllables for the virtual choir to sing (see Figure 14.3) but each audio instance was recorded individually and then mixed together for the interest of variety as slightly different samples would be triggered from the MIDI with each take due to the round-robin sample playback. The lyrics were written by me in ancient Greek and were loosely based on extracts taken from my old high school textbook of Homer's *Odyssey*. Sometimes, gibberish words were chosen purely for their acoustic resonance rather than their meaning but there is a clear reference to the story of the

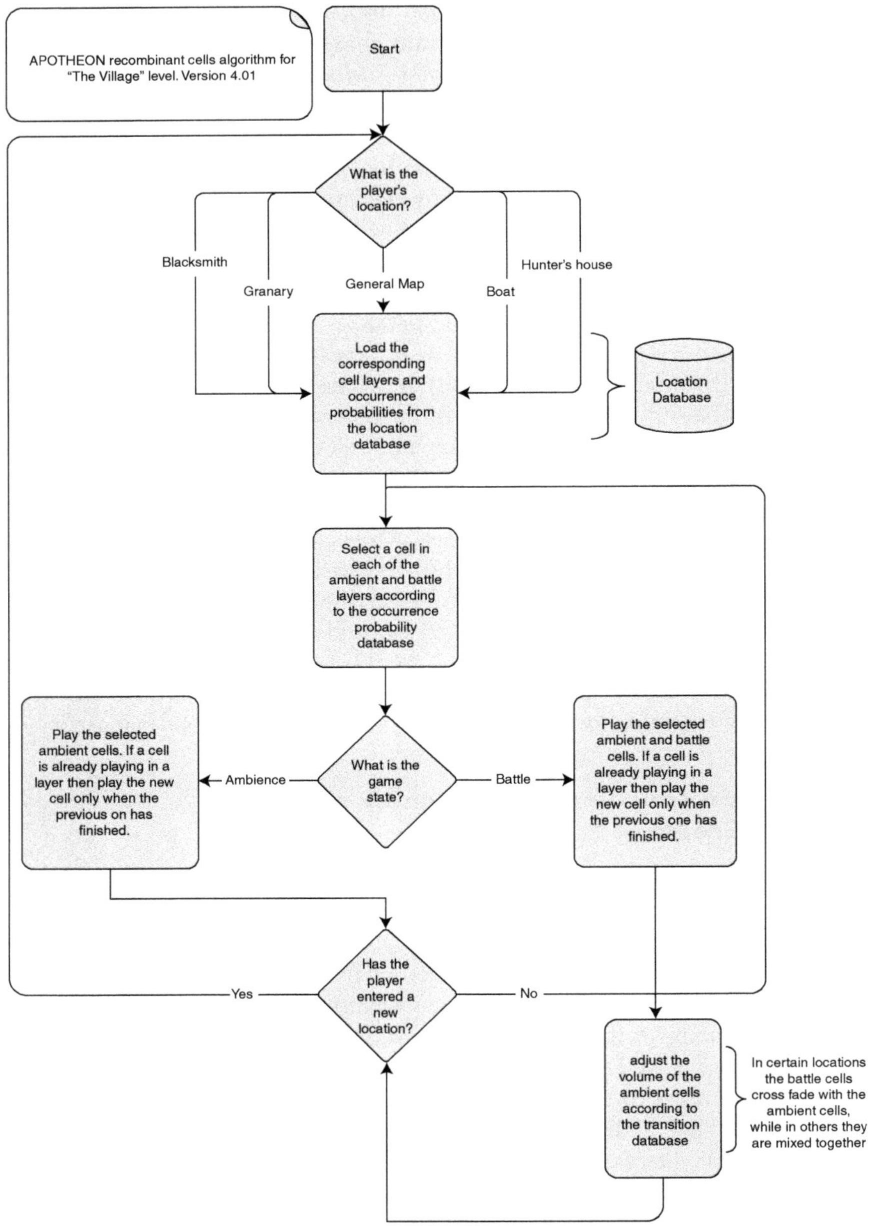

Figure 14.2 A Flow Chart of the recombinant cells system for *The Village of Dion* area.

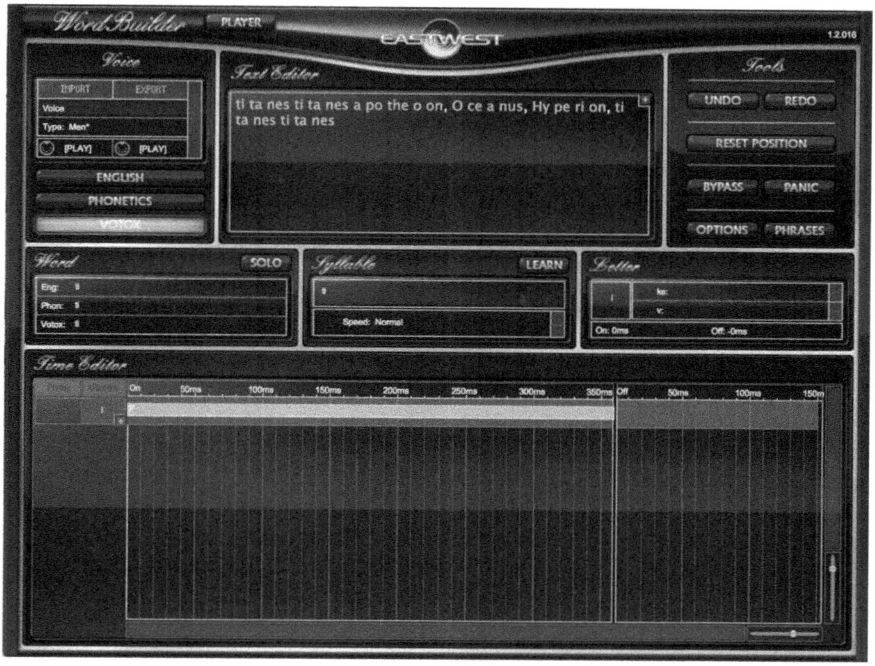

Figure 14.3 Parts of the ancient Greek text sung by the virtual choir in *King of the Gods*, written in separate syllables using East West's *WordBuilder* software.

Titans, who were the previous rulers of the world before being overthrown by Zeus. The composition is structured around two themes; the main theme in 7/4 and a secondary theme that follows in 7/8.

Takeaway tasks

Task 1 – Algorithmic composition (very challenging) – Create your own recombinant cells system

You can give this technique a try by creating a simple composition of three layers organized according to function (ex: melody, accompaniment, percussion) that each has approximately five phrases. You can assign each phrase with a probability percentage (it does not have to be an equal % among them), and make sure to include a probability for silence to occur in each layer so you do not always have the same number of instruments. Some tips include to use proportional lengths for each phrase (half or double) if you want the rhythm to remain in sync, and a harmonic approach that would work in most combinations (ex: most cells could be following the same chord progression, but some rare cell combos could be more dissonant or break away from this pattern). There will probably be a lot of trial and error needed before you are satisfied with the musical output, but the beauty of this idea is that once it works well it will generate a nearly endless number of variations. You can read the next chapter on *No Man's Sky* to get more ideas in this area!

Notes

1 Banas, "Soundtrack of the Decade: Honorable Mentions That Mustn't Be Ignored."
2 Aristopoulos, "A Portfolio of Recombinant Compositions for the Videogame Apotheon". https://openaccess.city.ac.uk/id/eprint/19298/.

Bibliography

Aristopoulos, Marios. "A Portfolio of Recombinant Compositions for the Videogame Apotheon". 2017. https://openaccess.city.ac.uk/id/eprint/19298/.
Banas, Graham. "Soundtrack of the Decade: Honorable Mentions That Mustn't Be Ignored". *Push Square*, 2020. https://www.pushsquare.com/news/2020/01/soundtrack_of_the_decade_honorable_mentions_that_mustnt_be_ignored.

Chapter 15

No Man's Sky (2016)

A conversation with the audio director Paul Weir

About Paul Weir

Paul Weir is an audio director, composer, and sound designer who has worked in over 40 games including *Discworld II*, *Lego Batman*, the *Last Campfire*, and *No Man's Sky*. Aside from being the audio director of *Hello Games* he runs *Earcom*, an audio production company with a specialty in generative audio.

About the game

No Man' Sky is an exploration and survival game that lets players embark freely on an epic voyage set in an infinite procedurally generated universe (video example 62). Despite the relatively small development team of *Hello Games* the game received massive publicity and has constantly been updated since its launch in 2016 (Figure 15.1). The music was composed by the English post-rock band *65daysofstatic*.

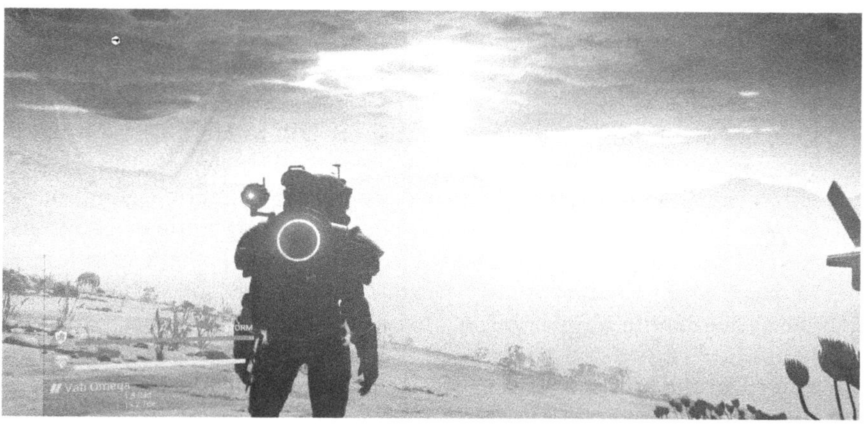

Figure 15.1 A gameplay screenshot from *No Man's Sky*. The flora and fauna ecosystem of each planet is procedurally generated and there are over 18 quintillion possible planets to explore.

DOI: 10.4324/9781003146872-16

Composition technique – Generating music for an infinite universe

MARIOS ARISTOPOULOS (MA): Would you describe the music system in No Mans Sky as generative or procedural? What is the difference between these terms?

PAUL WEIR (PW): Most of my work, although it gets called procedural, I would say it is generative. For me, that is dynamically arranging and creating music. It is taking elements, often at a micro level, and decide what to play and how to play it. What procedural music does instead, is to use a musical algorithm to define musical terms and phrases and build brand new elements of music. I see it at as a scale, from linear composition on one end of the scale to procedural on the other. I have done everything and often the most effective approach is a combination of different techniques. What it comes down to is as a composer I am trying to achieve a certain goal, or a set of goals. I want the player to feel something, and I am not interested in technology for technology's sake. *No Man's Sky* is not linear, there are no levels, I don't know what the player is going to do at some part of the game, so it makes enormous sense when the game is in free form to have free form music.

MA: Do you think that generative music systems can offer benefits beyond generating variation?

PW: For me it is a lot more than trying to prevent people from getting bored by generating some new variations. It is how do you wrap the musical experience around what the player is doing to create moments of synchronicity. In *No Man's Sky* if you behave in a certain way the music will change to reflect that, so it feels like it is a personalized experience. You are building all these bridges with the player.

Another good example of that which is not in games, comes from the generative systems I wrote for real world spaces, principally shopping malls, banks, places like that. These were not art installations; they were not meant to be noticed and they were completely commercial. One big one I did was the main transport hub in Helsinki and that is remarkably similar than working in a game, in that you are trying to enhance the environment, and enhance people's experience and make people feel something. The system could take any number of inputs, for example look at how busy the environment is through speech level or noise levels, and obviously season, time of day, or whatever it is, the system could absorb any of that and control the values. Maybe if people are rushed and stressed then just bring the mix down a little bit and calm the arrangement, or if it gets busier then you may introduce different elements, so it is much more of a kind of dialogue with the personal experience of the music.

MA: **Could you summarize how does the PULSE System work in *No Man's Sky*?**

PW: It is very simple in many ways. It sits on top of Wwise and it absorbs music content that we feed in to it, usually simple phrases or a couple of bars of a drum loop, or things like that, and we can then create what we call instruments out of that. For example, drones are very easy, here are 20 files, I am going to create a drone [video example 63]. You can attach behaviours to that: how often does this sound play, what scale can I play in, it might need to be re-pitched, or play it every X number of bars. If it is a drum loop, it is utterly important that every bar plays sequentially, and a lot of work was put into making sure that time keeping is very reliable. Then you can take collections of these instruments and move between different behaviours in a type of canvas. These behaviours would be connected to different sections of the game. For example, during space flights, gameplay parameters such as am I moving, am I facing a planet, am I heading towards a planet, and things like that would affect the music generation [video example 64]. As you move around the behaviour changes, very gradually and smoothly by morphing very nicely between different steps.

To a certain extent, I am not that bothered about what the behaviours are, it was really to solve a problem in that a lot of generative music, yes, it is kind of randomized, it does kind of adjust over time, but it never actually changes, as it is using the same rules. If you slice any point over time, it sounds the same. That is the limitation we also had with this commercial retail music. You want these gameplay drivers to keep changing the music, to keep forcing different behaviours.

MA: **Does the PULSE system develop the music only when there is a gameplay parameter change?**

PW: The system can do both, generate variations but also play this very specific music for a situation. There are multiple types of programming logic: it can do this AND that, this OR that, or if it is doing that then NOT do that. It is not a fantastically sophisticated system as we were a small team, 15 people, we spent a lot of money on creating procedural SFX, which are properly procedural, and we already had a lot of experience building this kind of music systems, and because we were working with *65 Days*. It was a nice middle space, where we could do something which we knew it will help the game, where *65 Days* felt comfortable with, but also that suited the game and suited our technical approach.

MA: **How did you ensure that there will be a sense of musical coherence and meaning with the system?**

PW: That is partly done by *65 Days* in their music. We did the sensical theme, first of all letting them compose in a very traditional way, letting them basically write an album [video example 65]. But because they are very technical as a band, alongside that, I wrote a set of rules and explained how I was going to use it. Not all composers feel comfortable writing this way, some think of the implementation process too much rather than what it is they are trying to compose. We are quite careful about that, do not worry about technology, I know what I need from you, because I have done it before. Write what you think is effective music. That is where the cohesion and meaning came from.

MA: How much were the composers, *65 Days of Static*, involved with the technical side of the generative system?

PW: *65 Days* didn't use PULSE, they just had this set of rules and then they would write me a text list of what should go where in the game, and I would interpret that into the system. There are two things to avoid. You give a composer a set of rules and they compose to the rules and they kind of lose their sense of composition. You do not want that. The other thing you do not want to do, and I have seen both in my career, is to try and repurpose existing music in a kind of generative way. Probably it is not going to work unless you do stem mixing which for me has nothing to do with generative music [see Chapter 12 Vertical Layers]. The middle ground is nudging people carry on doing what they are great at doing but guide them to give you the technical elements you need. For example, I really like those 4 bars can you give me more variations on these? Do not worry about time, we 'll fix all of that. Setting some light rules and then separate the music that you need to feed the system to make it work. The rules can be very simple (ex: key, tempo, number of bars), and the outcome can be very complex. If you flip it the other way, if your rules are overly complex, you are just creating barriers all the time. *No Man's Sky* is in many ways a very complicated score, but each individual element is really simple.

MA: When I created my generative score for the game *Apotheon*, I struggled with creating a real sense of development for longer section. After a while, I felt the music gets "predictably unpredictable". What was your experience with this?

PW: This is the exact feedback we got on NMS early on, exactly that, "it is great, but it always sounds the same", that is why a lot of the work I have done is on how to drive change in the system, and again that can be really simple, but inevitably you need inputs from the game. Whatever it is just choose some inputs, you can do it in Wwise with some RTCPs or even just don't have generative music going for too long. We did a game called *The Last Campfire*, that is mostly linear music but there is some generative music in the larger areas which none picks out, it has never been commented on! That is because it is super simple, every piece, could have random elements of music in there with time, beats, key, which generates a nice evolution over time by just adding variations to the linear elements.

MA: Where do you see the future of generative game music for the next decade?

PW: Obviously, people are already using Machine Learning for music generation. That is a whole other conversation, a separate conversation to have. The pure machine learning driven compositions I have heard, are perfectly functional but not very interesting to me. It has clearly evolved but it is not where I sit, that is not my job, it is someone else's job doing all that. I think where Machine Learning could be really interesting is expanding music that you create, feeding it into a system, understanding what you are doing, and generating more music that is building on that.

Another area that I am interested is how do you develop your music based on player input in a way that feels musically satisfying and effective for the player in a more intelligent way. That could be going towards more, using a lot more metadata to describe what each musical element is, and to describe what behaviours could be attached to it. Almost to create chunks of logic. For example, this bit of logic is great at making chords that make you feel awesome, attaching that as a behaviour and having a cloud of behaviours. This has nothing to do with ML, it is all scripted in a way. Rather than having a bunch of phrases and randomize these, you are pulling behaviours on top of elements of music and combining them.

Production tools – Software for getting started with generative music

MA: Which software would you recommend for someone who wants to start getting involved with generative music?

PW: Wwise could be really good for that. You can use random containers, you immediately have enough technology there to start experimenting and playing with ideas. Also, it is not hard to mock up these things yourself using any sequencer like Reaper or PreSonus Studio 1 that I use. You can create a bunch of phrases, throw them into Reaper, and tell Reaper to randomize their placement, see how it feels, that is an easy way to start. I don't use Ableton Live, but I know it can be useful in this area. Reaktor is also great but it is a little bit more complicated.

Sometimes in generative systems there is this tendency to be a bit gentle a bit ambient or harmonically very simple, so everything fits together. You should aim to be harsh on yourself, in a positive way. If your system only works if you make the music exactly in a certain way, then push it and evolve it. Throw stuff all over the place and test to see if your system is still working. Think about what you can do to make it more challenging, I always try to do that in my career.

MA: I remember in your GDC lecture, you mentioned that a good generative system does not need to be incredibly technical.

PW: No, it really doesn't. I always come up to you are a composer compose, how can you best service the interests of the game. A lot of coders will make sometimes overly complex systems to prove how cool their code is. I have been fortunate to work with many amazing coders, such as Sandy White, who built procedural audio systems and helped build PULSE, what he said about programming, for him, he has been going up for 50 years, programming is not what you type into a computer, programming is what you are writing pen and paper and designing a system, that is all. The creative aspect of it, that is programming. Everything else is figuring out the technical details.

Career tips from an audio director

MA: Do you have any advice for student composers?

PW: I think a lot of it is mindset, with composers. As always, listening to a wide range of music, I am self-taught, being free, experimenting, not feeling constrained, I should be able to do this but I can't do this, fine, just have a go! Explore and have fun but at the same time have a critical creative brain, that is the essential bit in of all the work we do. Your internal critic, in a positive way. Is this working, is it not working, what is most effective, what can I try, how can I push it, how can I take elements of music I have already listened to and feed that in. I always say to my students, you need to be enthusiastic, you need to know about the industry, but you also need to be skilled. But if you are good at what you do and you are enthusiastic and you have knowledge then you will get a job. It is true for all creative arts.

MA: I find that this is true for in-house audio design jobs but for composition jobs I think that the game industry can be a brutally competitive environment!

PW: I acknowledge that I am fortunate, I do not look for work. For a while I worked mainly in sound design, but for most of my career it has been a 50/50 balance between composition and sound design. It is brutal but there are paths into the industry and again it is a creative industry, if people who are established listen to a student and their work is exceptional, they will acknowledge and support them.

MA: **Do you have any advice for how composers can reach the right people? Audio directors usually have little time to check an influx of music demos.**

PW: The path is still the same as when I started which is get to know people. You obviously need to have a strong portfolio, but you almost certainly will not get work only on the back of it. It will be meeting and connecting and then being able to back that up with your work. Other than that, it is the usual: do your game jam, get to know people in the same level, that is a very effective way, absolutely. I see almost everyone go through that pathway now. The problem with sending music to audio directors or composers, if for example someone sends music to me, is that I will listen to it but I do not know what to do with it because I am a composer so I am not going to give my work to a student. Some composers have more of a factory approach, but I don't work that way. So, I am not the right person to give it to, you are asking work from me but that is my work so why would I give it to you!

Having said that, that is how I got a job in the industry. I was very lucky, I was not looking for a job. I was going to a completely different job interview but met the audio director and he said well we are expanding at the moment so he listened to my music, and he literally was like "yes your music sounds good, we have loads of games, I will give you a job". You can have that lucky break. But, it is a human connection. There is a composer I spoke to a while ago, she was an excellent piano player, I direct her to go to the right places, go to the conferences, go to the meet-ups, do the game jams, she is getting work in the industry now. I think that is where other composers can help, you might not have my work, but I can point you at the right direction for who you should be talking to and support you.

MA: **Fantastic, thank you Paul!**

Takeaway tasks

Task 1 – Generative composition (variable difficulty) – Create a simple generative piece inspired by the techniques discussed in this interview

You can set your own rules as you like but a good starting point is to have a fixed key and tempo. You can test your system in any software you are familiar with. For example, in Ableton Live you can explore the nine different "Follow Actions" that control what happens after a clip finishes playing, in UE5 you can use sound cues that allow many options for randomization through visual scripting, and in Wwise you can use playlists and random containers. Once you get something working try to push the system further to produce output that is unexpected but also remains musically interesting.

Chapter 16

Doom (2016)

The Doom Instrument – Using FX chains creatively

About the game

This is a reboot of the all-time classic *Doom* franchise that helped establish the genre of First-Person Shooters in 1993. You take the role of Doom Slayer, a space marine on planet Mars that must overcome endless hordes of demons using any means necessary (from chainsaws to plasma blasters!). The game has received overwhelmingly positive reviews from fans primarily for its fast-paced action, brutally violent 3D graphics, and adrenaline pumping soundtrack.[1]

Fun trivia

Doom games have a history of containing numerous easter eggs left for players to find and five of them have been discovered so far within this game's soundtrack. Running specific parts of the audio through a spectrogram or tempering with the tempo and pitch will reveal hidden voice messages and diabolical images meant as a joke by the composer.[2]

How did the composer get the gig?

Mick Gordon began his composing and sound designing career by sending demos to developers in the early 2000s. After gathering some interesting credits under his belt (*Need for Speed*, *Marvel Super Hero Squad*) his career really took off with the music for season 1 and 2 of the fighting game *Killer Instinct* in 2013. He was then asked to compose the music for another iconic FPS franchise, *Wolfenstein: The New Order* in 2014, developed by the same company as *Doom* in 2016. After the enormous success of the *Doom* soundtrack, he moved on to compose the music for the sequel *Doom Eternal* released in 2020, but sadly, according to recent interviews from both sides he might not be continuing with the franchise after disagreements on the mixing of the accompanying soundtrack.[3]

DOI: 10.4324/9781003146872-17

Production 101 – Signal flow

To understand how the Doom Instrument works it is important to first understand the fundamentals of audio signal flow. Signal flow is a term that describes the path an audio signal takes from its original source to its final output. It is a simple yet powerful concept that can cause a lot of wasted time and unsatisfying mixing results if it is not applied correctly. As a metaphor, you might find it helpful to think of water flowing from a lake through a complex system to reach your tap. By using valves and pipes you can control the flow of water as you wish but if the flow within a pipe is interrupted it will not reach its destination. Similarly, if you cannot hear any sound from an audio system, the easiest way to troubleshoot the problem is to track the signal flow starting from the source (ex: a microphone) and then testing if audio is present at each stopover (ex: an FX unit) until you find what is blocking the flow (ex: a muted channel, a broken cable, a fader, lack of power, etc.).

When Effect Units are incorporated into an audio system, they are usually connected so the audio signal will flow in **Series** and/or in **Parallel.** In serial, the signal will go through each FX unit one after the other, adding up any processing along the way. In parallel, the signal will split to a new path that will be unaffected by any processing that happens beyond that point in the previous path, but the two paths can be recombined at a later stage. For example, in a DAW the audio effect slots within a channel are connected in series, while using a send to an Aux track is a parallel connection.

Composition technique – Creating The Doom Instrument with sine waves and FX

Inspiration for The Doom Instrument

The audio team of *id software* has a music tradition called *Weird Wednesdays* in which members jam with each other with complete creative freedom to encourage new ideas that have not been tried before. While exploring possible musical directions for the game, composer Mick Gordon found inspiration in the game's concept of *Argent Energy*.[4] This is a new form of renewable energy that is powering the entire world and it is controlled by an evil corporation that supposedly mines it from the planet Mars, while in secret it is taking it directly from hell!

To mirror the idea of pure energy running through a futuristic power grid (Figure 16.1), Mick came up with *The Doom Instrument*, a system that

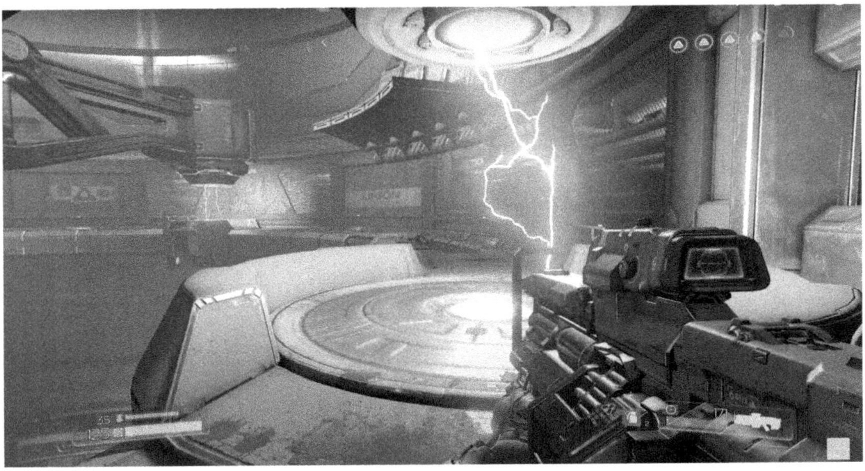

Figure 16.1 A gameplay screenshot from *Doom* showing the flow of *Argent Energy* that inspired Mike Gordon's concept for the creation of his Doom Instrument.

starts with loops of low frequency sine waves (the purest form of sound) that run through multiple arrays of analogue high-end audio processing units.[5] This instrument produces harmonically complex textures that are constantly evolving and it became one of the core features throughout the *Doom* soundtrack along the massive metal guitars and drums (video example 66).

Signal flow in the Doom Instrument

The system begins with an input of sine wave riffs that loop within the subwoofer range. The signal is immediately split into four parallel FX chains, each serving a different purpose.[7] As you can see in Figure 16.2, the first two paths can add numerous types of different distortion and therefore harmonic content to the sine waves. Some of the pedals might change but the concept remains the same. The third path is primarily adding reverb and delay through vintage tape machines, and the fourth is adding real time feedback by playing the audio through a mini guitar amp that is using a live microphone within another closed feedback loop.

The third and fourth pathways can also serve as additional sources of light distortion through the guitar amp or by simply over boosting the signal beyond what each circuit can handle before being distorted. If all the effects were used in series (one after the other) like in your typical guitar pedal setup, the output would quickly become a blurry mess as each effect

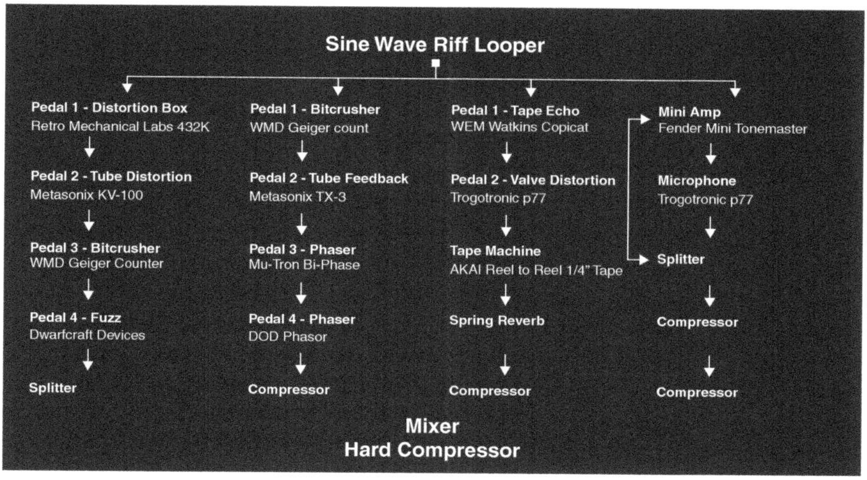

Figure 16.2 A chart showing the signal flow in the Doom Instrument along all the specific pedals that Gordon used based on his 2015 GDC lecture.[6]

would build upon the sound of the previous one in the chain. Just imagine having eight distortion pedals along two tape echos, reverb, and constant feedback on the same signal! However, by splitting the sine waves into four independent paths and by adding further splitters along the way at key points (ex: at the end of the first path and in the middle of the fourth path) the composer wields a much tighter control of the sonic output. For example, he can add reverb to the pure sine waves but not add it to the distorted ones, or he can create feedback loops only between specific FX units.

Adding chaos into the system

What makes this system particularly playful is not just the exotic and expensive collection of guitar pedal candy, but also the chaotic textures that it produces as a response to simple changes in signal amplitude. This is achieved by using several gates and hard compressors at key points. In case you are not very familiar with compressor and gates: A compressor reduces the dynamic range by making loud and quiet waves more similar, while a gate can open and close based on a given loudness threshold and thus stopping or allowing quiet sounds of going through a signal chain. Gates are typically used in music production for blocking out unwanted noise by shutting down a signal when there is not much going on (ex: to take out the guitar buzz). However, they can also be used more imaginatively. In the iconic track *Heroes* that was inspirational to Gordon, David Bowie used three different reverbs placed in parallel, each with their own gate that would block the signal at ascending threshold levels. This allowed Bowie to dynamically control the amount of reverb in real time only with the power of his voice, as calmer moments would sound closer while louder dynamics would open more of the reverb gates (video example 67).[8]

In the Doom Instrument, the presence of multiple gates with different attack, release, and hold times (how long the gate remains open), introduces unpredictability into the system as sine waves of different amplitudes might open and close different gates at different times. Therefore, the composer can sit down and play this system as an instrument by manipulating a parameter on one FX unit and observing the impact on the rest of the system. According to Gordon at some point: "the machines start doing things on their own rather than you sitting down and using the machine".

At the final stage of the signal flow the signal of each path is mixed together, and a hard compressor (with a very high ratio such as 20:1) puts the cherry on the cake by altering the sound in two distinguishable ways that are evident across the soundtrack: First, by having a longer attack time, the hard compressor allows the aggressive and loud staccato riffs to go through, and completely crushes everything that comes after. This on/off punchy approach was a conscious choice by the composer to ensure that at least some of the music would be able to cut through the very hectic SFX mix from the battles. It is interesting to note that this is another creative solution to the same game audio problem of music clashing with action SFX mentioned in Chapter 1: *Space Invaders*. Second, by having a very quick, almost instantaneous attack time and a very long release time the hard compressor completely crashes the loud sounds but gradually allows all the intricate details of the electronics of the Doom Instrument to swell in and produce the beautiful screeching type of effects that give this soundtrack another one of its signature sounds (ex: video example 66, 24:20–25:50).

Production tools – The doom guitar sound and using a Shepard Tone

The doom guitar sound

Doom has some of the heaviest metal guitar sounds you will find in any video game. The amusing part is that the developers initially clearly specified in the brief that the music should not have any guitars at all but thankfully Gordon was able to change their minds. To achieve this massive wall of guitar sound he followed three steps:

1) He started by drop tunning a 7th string electric guitar but as the result did not feel low enough, he used a 9th string electric guitar instead![9]

2) He used a technique inspired by Marilyn Manson in which you record the guitars an octave higher and in double speed, but then you pitch shift them down and slow them to the original tempo by using an analogue tape machine playing in half-speed. According to Gordon this might not work as well if it is emulated on virtual tape plug-ins as you need a slow playing tape to get the natural saturation and harmonic distortion needed.[10]
3) A secret ingredient that is used occasionally in the soundtrack is the use of an audio **morphing** plug-in called *Morph* that uses different algorithms to combine the guitar sounds with different SFX from the game (ex: a chainsaw recording) and transform them into a new hybrid sound that shares sonic characteristics from both.[11] To make the best out of this technique, it helps if the two sounds are set in the same key.

Shepard Tone

The demon slaying cinematics that trigger in the game (known as glory kills) have a varying length, so Gordon created a number of risers using a Shepard Tone. A **Shepard tone** is a fascinating aural illusion in which a series of tones can appear to rise (or fall) indefinitely, and it is produced by a superposition of sine waves that are altered in volume and are separated by octaves or major sevenths.[12] It is named after the cognitive scientist Roger Shepard who invented the technique and has been creatively utilized by many media composers: in *Super Mario 64* the music rises infinitely to mirror climbing an endless staircase (video example 68), in the film *Dunkirk* it was used extensively by Hans Zimmer to create a constant feeling of tension, while in another Christopher Nolan film, *The Dark Knight*, Shepard patterns were used on real vehicle recordings to create the hypersonic sound of the Batpod (video example 69).[13]

Takeaway tasks

Task 1 – Production (very challenging) – Create your own Doom Instrument

You do not have to necessarily rely on expensive guitar pedals or analogue gear, you can replicate this in your DAW using AUX tracks, buses, and stock plug-ins. You can also swap any of the effects to whatever you prefer as long as the concept remains the same: you use parallel FX arrays with gates and compressors along the paths to dynamically respond to changes in volume and add some chaos into your textures. You could start this with two arrays and then expand the system to your liking. Remember that tweaking the attack and release of a hard compressor (over 20:1 ratio) at the end of the chain can make all the difference in determining how the sound will be shaped.

Task 2 – Production (challenging) – Creating an infinite riser Shepard tone

There are multiple ways to create a Shepard tone illusion. I recommend starting with two sine waves or two notes on a simple instrument that are placed an octave apart and either ascend or descend along the chromatic scale at the same speed. The trick is to make a perfect loop in which the ending of each sequence feels like the beginning of the next. To make the illusion more effective you can start with staccato notes and adjust the volume (or MIDI velocity) of the first sequence so it eventually matches the exact volume of the 2nd sequence. As you cannot obviously keep increasing the volume indefinitely you might need to fade out to match the volume of the beginning of the 3rd sequence. You can then just copy/paste this pattern across multiple octaves and that should be enough to do the trick! Have a look at how this technique works using a MIDI vibraphone in Super Mario 64 music (video example 70).

Task 3 – Production (challenging) – Recreate the Doom guitar sound

If you have access to a guitar, you can follow the three steps described earlier using substitutions as needed. A great plug-in that you can explore for audio morphing is *Tone Transfer* made by magenta and Google Research that uses machine learning to create hybrid sounds from two different samples. I also recommend using Xfer Records *OTT*, a free re-creation of a very aggressive multiband compressor. This plug-in is usually popular with dubstep producers, but it can also be used creatively to make any sound feel extremely loud and punchy.

Notes

1 "DOOM on Steam."
2 Ruiz, "Doom Soundtrack's Final Easter Egg Found Two Years after Release."
3 Wojnar, "DOOM Eternal Devs Say They'll No Longer Work with Composer Mick Gordon."
4 Gordon, *DOOM: Behind the Music Part 2.*
5 Gordon, *DOOM: Behind the Music – GDC.*
6 Gordon, *DOOM: Behind the Music – GDC.*
7 Gordon, *DOOM: Behind the Music – GDC.*
8 Gordon, *DOOM: Behind the Music – GDC.*
9 Gordon, *DOOM: Behind the Music Part 1.*
10 Gordon, *DOOM: Behind the Music – GDC.*
11 Gordon, *Mick Gordon Interview – Warren Huart: Produce Like a Pro.*
12 Shepard, "Circularity in Judgments of Relative Pitch."
13 Malinverno, "The Shepard Tone: What It Is and How It Works."

Bibliography

"DOOM on Steam". *Store.Steampowered.Com*. Accessed 5 September 2022. https://store.steampowered.com/app/379720/DOOM/.

Gordon, Mick. *Doom Music Panel with Mick Gordon – PAX AUS 2016*. Video, 2016. https://www.youtube.com/watch?v=Pu4dB_Wy1-E&t=1571s&ab_channel=GrandJasonGaming.

Gordon, Mick. *DOOM: Behind the Music – GDC*. Video, 2017. https://www.youtube.com/watch?v=U4FNBMZsqrY&ab_channel=GDC.

Gordon, Mick. *DOOM: Behind the Music Part 1*. Video, 2016. https://www.youtube.com/watch?v=ua-f0ypVbPA&ab_channel=MickGordon.

Gordon, Mick. *DOOM: Behind the Music Part 2*. Video, 2016. https://www.youtube.com/watch?v=1g-7-dFXOUU&ab_channel=MickGordon.

Gordon, Mick. *Mick Gordon Interview – Warren Huart: Produce Like A Pro*. Video, 2016. https://www.youtube.com/watch?v=-bsXuaIVMB4&ab_channel=ProduceLikeAPro.

Malinverno, Matteo. "The Shepard Tone: What It Is and How It Works". *Splice.Com*, 2022. https://splice.com/blog/how-shepard-tone-works/.

Ruiz, Michael. "Doom Soundtrack's Final Easter Egg Found Two Years after Release". *Dualshockers*, 2019. https://www.dualshockers.com/doom-soundtrack-easter-egg-final/.

Shepard, Roger N. "Circularity in Judgments of Relative Pitch". *The Journal of the Acoustical Society of America* 36, no. 12 (1964): 2346–2353. doi:10.1121/1.1919362.

Wojnar, Zak. "DOOM Eternal Devs Say They'll No Longer Work with Composer Mick Gordon". *Screenrant*, 2020. https://screenrant.com/doom-eternal-devs-composer-mick-gordon-controversy/.

Chapter 17

Call of Duty: WWII (2017)

A conversation with the composer Wilbert Roget, II

About the composer

Willbert Roget, II is a veteran composer that started his career in the game industry as a staff composer for LucasArts where he worked on titles such as *Star Wars: The Old Republic* (Figure 17.1). He later became a freelance composer scoring multiple AAA games such as *Mortal Kombat 11*, *Lara Croft and the Temple of Osiris*, *Destiny* 2, and *Call of Duty: WWII.*

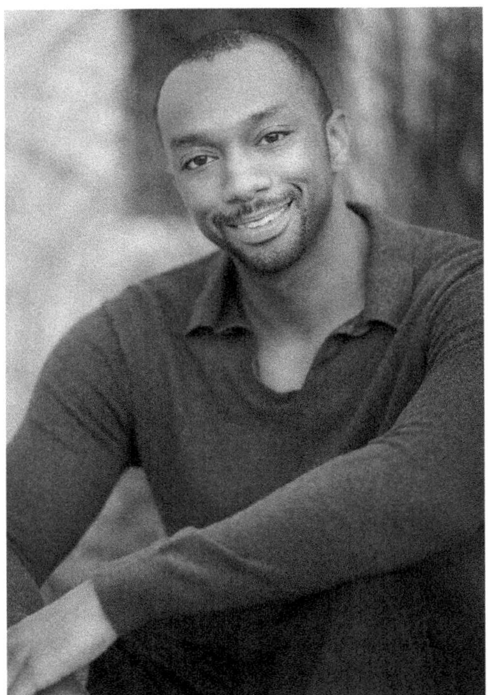

Figure 17.1 A photo of game composer Wilbert Roget, II.

DOI: 10.4324/9781003146872-18

About the game

Call of Duty: WWII is a first-person shooter game set in World War II. It is the fourteenth installment in the *Call of Duty* series, one of the best-selling gaming franchises that have sold over 400 million copies. Players land in Normandy on D-Day and experience the horrors and braveries of intense combat in a series of campaign missions across historic European locations.

Composition techniques – Synchronization and competing with SFX

MARIOS ARISTOPOULOS (MA): **One area that I found incredibly well done in WWII is how tightly synchronized the music feels in relation to the ongoing action. How did you (and perhaps the audio programmers) achieve such a cinematic precision to the development of the music?**

WILBERT ROGET, II (WR, II): The Sony PlayStation music team handled implementation for *Call of Duty: WWII*, which was a brilliant collaboration given that their studio is practically across the street from Sledgehammer Games! For this title, they generally opted for a detailed linear scripting approach, rather than systemic dynamic music. Every major moment in the campaign – for instance, defeating a group of enemies to move on to the next area, or triggering a cutscene – is accompanied by a custom trigger that plays a music cue specifically edited for that moment [video example 71, example of a trigger at 17:30]. We also had a very basic system for stealth music, playing stingers as enemies become aware of the player, and triggering action music if they engage in combat.

MA: **Relating to the previous question, the horizontal music transitions and stingers that appear to be triggered by gameplay events feel very natural and musically coherent. How did you plan this from a compositional perspective?**

WR, II: When writing game scores with Sony PlayStation, their music supervision team usually asks composers to write through-composed "suites" of music rather than specific individual game-ready loops. These are typically a few minutes long, beginning with an intro and having a natural musical progression through the different moods requested, and ending naturally as well. We deliver in a few dozen stems, and record the orchestras with deep striping, so that the music editing teams have as much material as possible to work from when cutting together cues for in-game use.

MA: **Another area that I felt you have approached very skillfully is how well the music works in relation to the intense SFX coming from the battlefield (see Figure 17.2). This is a common challenge for game composers in action games, could you please give us some insights of your approach?**

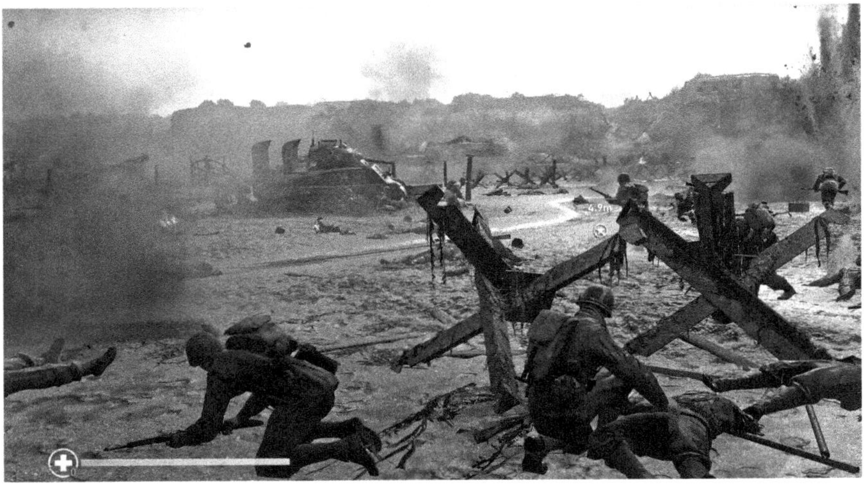

Figure 17.2 A gameplay screenshot from *Call of Duty: WWII*. Intense combat scenes such as this are usually accompanied by loud battle SFX which makes it challenging for the music to cut through.

WR, II: Several months before I was hired, the first conversation I had with our audio director Dave Swenson was about this exact problem. On a previous *Call of Duty* title, he had to mix a very dense, bombastic score against dense, bombastic sound design – both occupying the same frequency ranges with punchy and impressive high-tech sounds. Unfortunately, this meant that the music had to suffer in the mix, as gameplay-relevant SFX needed to take priority.

My solution for *Call of Duty: WWII* was to remove elements of typical action-score orchestration that could potentially clash with the game's sound design: There are no trumpets, high woodwinds, or mallet instruments, nor are there any snare drums or other percussion with very sharp transients. I also avoided writing particularly high parts for the violin section, loud traileresque action drums, or overt synthesizer parts. I then used solo strings and string quartet in most of the action cues for extra rhythm [video example 72, from 27:00], as well as extensive musical sound design based on processed recordings of WWII weaponry and vehicles to give a hazy "fog of war" vibe.

With all these restrictions and tweaks to the instrumentation, I ended up with a sound that could easily blend with the in-game sound design without the need for intense ducking. As a final check before delivering cues, I would play my in-progress music against video clips from particularly busy levels from previous WWII-era *Call of Duty* games – if anything poked out of the mix, or was completely masked, I would remove it.

Production tools – MIDI orchestration

MA: The orchestrations in the game sound very powerful and expressive. Unfortunately, most smaller games rarely have the budget for recording a live orchestra. Do you have any tips on how to get a similar orchestral aesthetic with MIDI instruments?

WR, II: This is a very complicated subject that I covered in my 2016 Game Developers Conference lecture, "AAA Virtual Orchestration On An Indie Budget". First, I'd recommend researching your favorite scores to find out where they were recorded, and model your sampled orchestra after that – not only the mix, but even your choice of what samples to buy in the first place. I had the opportunity to work with the London Symphony Orchestra at Abbey Road for my final *Star Wars* score at LucasArts, so I modelled my setup after that sound very specifically.

Next, I'd recommend working with reverb multi-dimensionally – having just one single reverb over the whole mix can create a flat, washy tone, so instead I use a hall reverb send, multi-mic samples, overhead as well as distant hall IR reverb sends for the samples that don't have multi-mic, and finally a subtle mastering reverb on the full mix.

For the IRs, I mix dry vs. wet send levels based on the individual instruments' overall loudness in real life, not on how "verby" I think it should be in the abstract. For example, a trombone is generally louder than a bassoon, so it would be louder in the distant hall microphone set, and thus I'd turn up its distant IR send levels. The bassoon might need more support in a live-recorded mix, so I'd turn up its dry signal or overhead IRs send levels. The idea is to use reverb "in reverse", pretending that the samples were recorded live and only allowing myself to use mixing techniques that would be possible for a live recording.

Once this is set up in my template, I don't allow myself to make broad changes to the mix, and force my mix problems to be solved with proper orchestration instead. For example, if a clarinet melody is inaudible, the solution is to thin out the accompaniment, double it with another instrument, or change the tessitura of the melody – I won't just reach for the volume slider or add EQ, compression, or other mixing effects, as tempting as it may be.

MA: I read in another interview that you wanted the orchestral mix to mirror the first-person perspective and focus on the protagonist's experience. What production techniques did you use to achieve this?

WR, II: The idea of "scoring in the first-person" isn't a production technique, but more of an overall mentality that you can bring into the composition. When I was scoring the unfortunately unreleased first-person shooter *Star Wars: First Assault*, I would load the game and fly through the levels, taking in the artwork and sound design, and imagining myself as a combatant. What specific emotions do I feel, and for how long? When a firefight breaks out, how do my emotions evolve over time? Those questions influenced my writing in terms of orchestration, harmony and especially form.

MA: I have seen this photo on your social media of a single Reaper Session that contained the entire soundtrack (Figure 17.3). Why did you prefer Reaper as your DAW of choice and why work within a single project session? You must have a supercomputer!

WR, II: Actually, I scored *Call of Duty: WWII*, *Guild Wars 2: Path of Fire*, and my *Lara Croft Temple of Osiris* score on a fairly modest 2012 Windows machine, with only about 32gb of RAM. Reaper is extremely CPU-efficient, project size doesn't really affect its performance at all, and I'm usually more efficient with my RAM usage than most composers due to my experiences in much older game music generations where we had limited amounts of RAM for samples during gameplay.

Keeping everything in one project file made starting new cues and revising old ones much faster, which was crucial with our unusually short deadline and high numbers of revision requests. It also let me easily copy recordings and musical sound design from cue to cue. With Reaper's render region system, each cue would export as though it was a separate file anyway – my orchestration and mixing teams don't see any difference between this and a typical setup. For the record, I only use the single-project method on scores that have a fairly consistent arrangement without too much variety in instrumentation or synth production – on Mortal

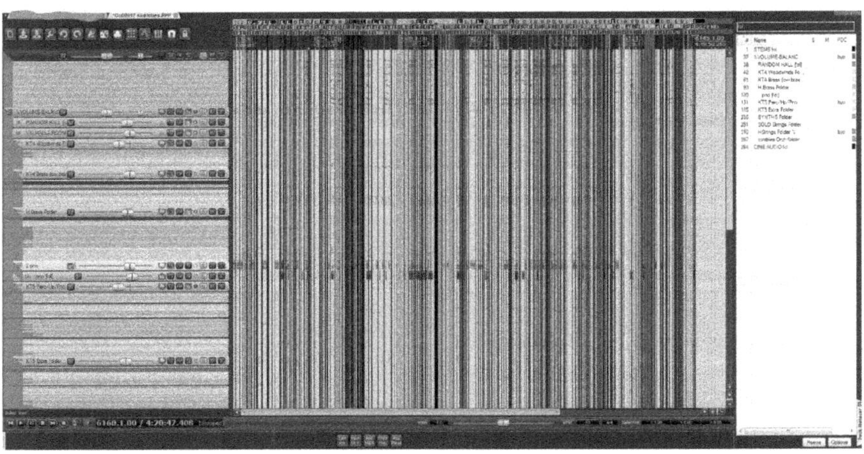

Figure 17.3 A screenshot from Wilbert's DAW session in *Reaper* containing all 4 hours and 20 minutes of the *Call of Duty: WWII* soundtrack in a single session.

Kombat 11 for instance, I used separate project files for everything since its instrumentation changes so dramatically from cue to cue.

Career tips from a AAA game composer

MA: The most common question from student composers is how to find work and network effectively with game studios. Any tips here?

WR, II: The most important advice I can give is to simply make friends in all aspects of the industry – other musicians, sound designers, programmers, QA testers, artists, designers. They all have fascinating stories to tell about their side of the craft, which can be greatly influential to your approach as a composer. Working in games is a very unique passion that isn't easily understood by people outside the industry, so it's important to have a balance of friends in and outside. As far as finding work, close friendships are what leads to gigs – not loose convention acquaintances or shotgun-method "networking". Spend your time with talented people you jive with, regardless of their position or status, and you'll grow a network of real friends that can eventually lead to more personal and effective introductions to the audio directors and game directors that might later hire you.

MA: Thank you so much for the interview!

Takeaway task

Task 1 – Production/arranging (challenging) – Recreate an orchestral recording of a game theme of your choice using only MIDI instruments

This task is harder than it seems as there are many details that need to be fine-tuned to achieve a realistic result. Apart from exploring Wilbert's MIDI tips from this chapter, having access to a good orchestral library that contains multiple dynamics, mic positions, and articulations can certainly make things easier from a production perspective. There are many fantastic options with high end professional libraries such as *Vienna Symphonic Library* and *Spitfire Audio,* but they can get very expensive. *East-West Composer Cloud* is an excellent subscription based option that you can turn on/off as needed, and another personal favorite of mine is *Symphobia* which is a little rigid but sounds great without any editing, so it is great when you are pressed for time. Choose a library that you personally like the sound of, as they all have different sonic characteristics and make sure to look for student discounts as most of the companies mentioned offer them. If you are on a limited budget there are also many free libraries available that can be very good to begin with such as *Spitfire Labs*, *BBC Symphony Orchestra Discover edition*, and *Studio Strings/Brass* if you are a Logic Pro user.

Chapter 18

Shadow of the Tomb Raider (2018)

Music as meditation, lost instruments, and 3D mixing

About the game

This is the 12th entry in the *Tomb Raider* series that follows the adventures of Lara Croft across numerous tropical settings in a thrilling race to save the world from a Mayan apocalypse.

Fun facts

The design team consulted with historians and locals to ensure that cultural depictions of the various indigenous civilizations portrayed in the game were accurate and respectful.[1]

How did the composer get the gig?

At the very beginning of the game's development, audio director Rob Bridgett started searching for composers who were specialists in South American music. Through a recommendation they found Brian D'Oliveira, a multi-instrumentalist, researcher, and composer from Trinidad and Tobago that was also based in Montreal Canada, a hotspot for game developers. Brian had previously composed additional music for games such as *Little Big Planet 3*, and *Resident Evil 7*.

Composition technique 1 – Getting into the zone/composition as meditation

Brian D'Oliveira chose to limit the instrumentation of the soundtrack to only acoustic sources, and to obtain as accurate representations as possible of many rare pre-Columbian instruments. He personally composed and performed all these instruments himself, with each take recorded in full, with no interruptions. There was no use of looping or editing of the material afterwards to make corrections and the performances were left intentionally imperfect and natural. His exploration of Pre-Columbian musical culture

DOI: 10.4324/9781003146872-19

also led him to experiment with other methods of music making practises that went beyond just implementing authentic instrumentation. He recalls:

> I had a major creative epiphany when I realised that expressing music from the viewpoint of the Pre-Columbian state of mind was accomplished with the understanding that all beings are intrinsically and unequivocally interconnected. Thus, it is a big reason why it's implicit in their ritual practices and daily lives and not seen as 'entertainment.' The deeper I went, the more my compositional methodology transformed, and I eventually reached a point of musical ease and transcendence where during the recordings I literally became a medium – without the need for thoughts or planning. So, towards the end of the game, composing for Shadow was mostly a matter of intent and then emotive expression. Often times it even felt as if the instruments and melodies were playing themselves, certain songs such as 'Return to Paititi' [video example 73] have an insane amount of fluid rhythmic complexity and non-tempered scales and textures that would have been impossible to create using a logical approach.[2]

The legendary film composer Vangelis was also a very strong advocate of such an improvizatory and meditative approach to music making. Vangelis wrote many of his iconic film scores by recording unedited performances in real time while watching the film on a projector. Although Vangelis made extended use of electronics, contrary to Brian's purely acoustic approach, the fundamental strategy that both composers shared is the immersion into a deeper exploration of improvizatory ideas that are captured in an uninterrupted process. It is important to clarify that, as it can be seen from Vangelis's interviews, some technical preparation and planning was a key element that took place before the recording would begin to allow the flow of ideas to remain uninterrupted. You can observe this preparatory process in video example 74, where Vangelis sets numerous foot controllers to be able to change his orchestration on the fly.

Implementation 101 – Music stingers

A **music stinger** is a common technique in interactive game composition in which a short musical phrase is triggered by a specific gameplay action. Stingers can be either **synchronous** or **asynchronous** in relation to the primary music track. Asynchronous stingers are extremely simple as they basically work like SFX: they can be triggered at any time and are unrelated to the timing of the rest of the music. For example, as we saw in the previous chapters on *Zelda* and *Mario Kart*, discovering a secret or receiving a power-up would result in a short musical motif that worked in the same way as a musical sound effect. On the contrary, synchronous stingers are aligned with the timing of the primary music and must be harmonically compatible.

Using audio implementation middleware software like *Wwise* can provide sophisticated options on the design of synchronous stingers. *Wwise* keeps track of the tempo and meter information of the music being played, and once a stinger is triggered the system can wait for a specific timing to play it so it will be perfectly synchronized with the rest of the music. As you can see from Figure 18.1, you have different timing options on when to fire the stinger such as immediately, at next beat, or at a specific cue timing that can be indicated by the lines within the wave editor. *Wwise* can also select among multiple stingers depending on the key, or even play pre-determined transition segments before them to ensure musical compatibility.

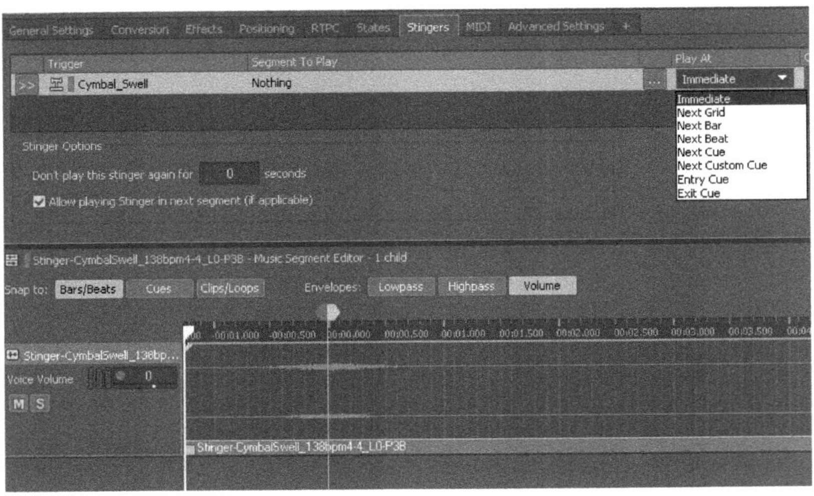

Figure 18.1 A screenshot from the stinger system within *Wwise* audio middleware software. The playback menu on the right and the audio editor in the bottom which allow multiple synchronization options.

Composition technique 2 – Adding interactivity with music stingers and music triggers

Despite the music being recorded linearly, the game features an impressively tight synchronization between gameplay action and musical accompaniment. This was achieved by handing over all the recorded stems to a team of audio implementation specialists who broke down and organized the music into smaller segments that were programmed to work interactively.[3] This process of composing music in a traditional linear fashion first, and then adding interactive elements later might appear strange for an interactive context, but it is common for bigger AAA games that can afford to hire both composers and audio programmers to focus on a different part of the process that fits their expertise. It is noteworthy that Wilbert Roget II in *Call of Duty: WWII* and the band *65daysofstatic* in *No Man's Sky* both followed a similar approach.

One of the primary techniques that the audio programmers used to achieve this in *Shadow of the Tomb Raider* is the addition of music stingers that are intertwined with the rest of the score in a natural way using both synchronous and asynchronous methods. The audio programmers did such an excellent job with editing and designing this system that these mechanisms might occasionally even pass unnoticed by players unless they are actively looking out for them. Have another look at video example 75 and observe how the music responds to sudden gameplay changes at 0:38, 0:45, and 1:10. It is obvious that during these unexpected moments that Lara almost falls to her doom, the music does not just happen to be perfectly synchronized to the action, but there is a trigger connected to each event. The music highlights and enhances these tense moments often instantaneously, or with minimal delay so it can synchronize with other animation triggers.

Aside from the good technical setup of these music stingers, this technique also works well because of the ambiguous rhythmic and tonal language of the compositions themselves, which make is easier for new elements to blend in the overall musical texture. For a stinger to be successful, it needs to balance between retaining musical unity with the rest of the underlying music and providing enough emphasis on the new event. Overall, the stingers in the game usually have one of the following elements:

1) fast and strong swells
2) sudden accents
3) new instruments
4) extended instrumental techniques (ex: flutter tonging, tremolo sul ponticello)
5) textures/sound design on the higher or lower extremes of the pitch registers

Composition technique 3 – Mixing the music within the 3D game world

Lastly, another interesting feature of the music that further immerses players deeper into their exploration of the ancient Aztec tombs, is a sophisticated mixing technique called **sound spatialization** (also known as sound localization). This term is achieved by a wide variety of audio technologies working together within a game engine to simulate the natural reproduction of sound phenomena in a 3D space in relation to the position of a player/listener. To achieve this simulation a game engine calculates numerous parameters in real-time such as the player orientation, the distance between the player and the sound object, any occlusion and obstruction by materials that might alter the sound, and the reverberation characteristics of that position.[4] Once all the parameters of sound spatialization have been set by the audio programmer/sound designer, the sound mix will be automatically adjusted by the game engine (or audio middleware) (Figure 18.2).

In contemporary games, the use of such spatialized sound techniques are very common in the implementation of environmental SFX. However, in *Tomb Raider* many of the Pre-Columbian musical instruments are also placed in 3D around the tombs using similar techniques. The practical implication of this is that as the player explores the tombs, the mix of these musical elements changes as their volume and panning adapt to the player movements. This approach discretely adds to the immersion as the

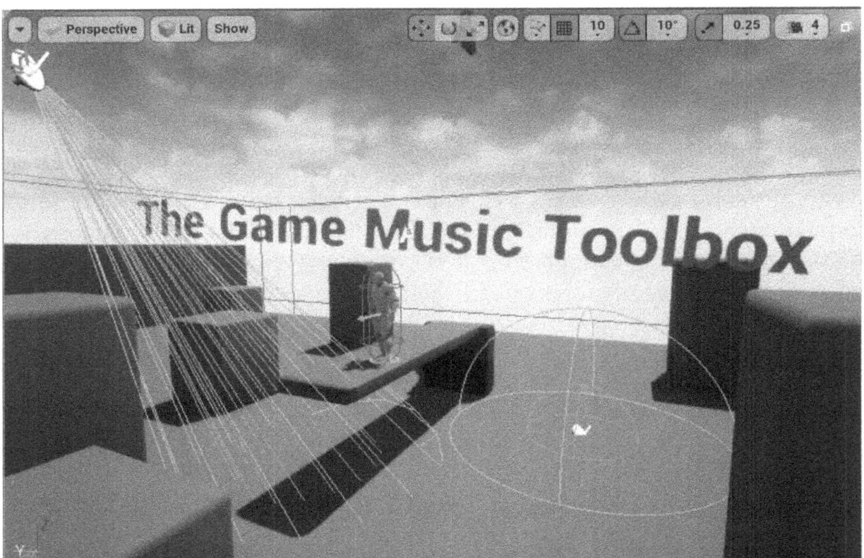

Figure 18.2 A screenshot of an imaginary *Tomb Raider* style environment recreated in UE5, to demonstrate how spatialized audio works. The sound on the left will only be heard while the character remains inside the cone radius. The sound on the right will fade-in once the character enters the outer sphere and play in full volume once he reaches the inner sphere.

music mix becomes a part of the tomb design itself rather than working as a fixed stereo image. The game was mixed in Dolby Atmos in Pinewood Studios in London but the spatialized effect also works with simple headphones.[5] You can observe it throughout all the tombs in video example 75 but it will feel more pronounced if you experience it interactively within the game.

Production tools – Hunting for lost instruments and the instrument sculpture

The setting of the game and its connection to the Pre-Columbian ancient civilizations was a major influence in Brian's choice of instrumentation:

> As Lara is in a much darker place emotionally, and a more dangerous place physically, the music needs to represent both of these things. The South American jungle and the Maya civilization both play a central role in the instrumentation of the score, and we are also reaching deeper into Lara's emotional point-of-view.[6]

To recreate these sounds, Brian went on an extended instrument-hunting trip around Mexico where he brought back eight bags full of 900 different instruments from small villages across the country.[7] He also spent time with local artisans and musicians to respectfully learn and understand how to play and customise the mechanics of many of them. Some of the unique instruments that can be heard in the score are volcanic rocks that were tuned in different pitches, clay skulls of various sizes, various wooden percussion instruments, and most important of all, an Aztec Instrument known as the Death Whistle, which is shaped like a puma but also sounds like a puma scream when you breath into it. According to Brian, the Death Whistle and the cello were the two primary instruments used to represent the two sides of Lara in the score, while various types of native flutes were used to represent the sound of birds from the Amazonian jungle.[8] You can see and listen to many of these amazing instruments in video example 76.

The instrument

Another truly unique and exceptional instrument that was also used in the game, primarily as a source of sound effects, is a custom commission from sculptor Matt McConnell shown in Figure 18.3. This instrument was made in collaboration with the composer Jason Graves that used it in the music of an earlier *Tomb Raider* game in 2013. There is an iOS app available where you can play a digital interactive version of it.

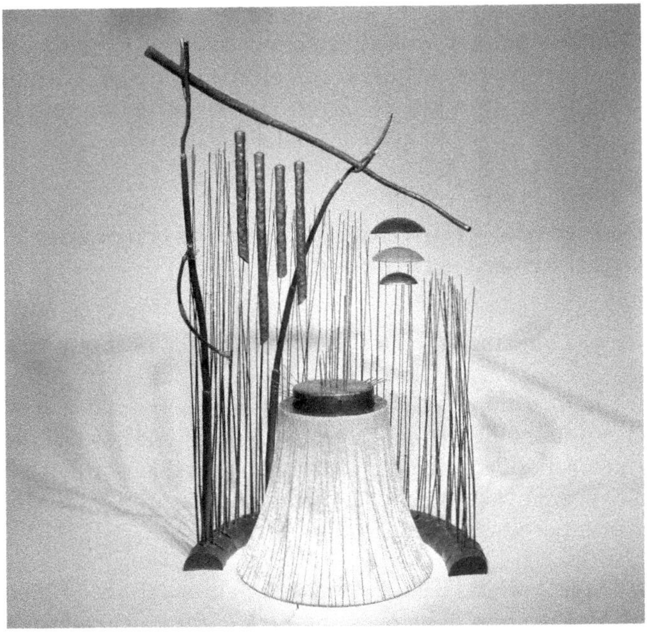

Figure 18.3 A photo of "The Instrument" sculpture created especially for the *Tomb Raider* games by Matt McConnell.[9]

Takeaway tasks

Task 1 – Composition/implementation (easy/medium) – Create a set of stingers for an imaginary level in Tomb Raider

Here are some suggestions but feel free to come up with your own events: animal attack, sudden drop, trap mechanism activated, danger averted, ancient treasure discovered. You can apply the same idea of musical SFX as in *Zelda* (see Chapter 3) but here you are not limited to using basic synth waves, the sky is the limit!

Task 2 – Composition (hard) – Compose music by recording an uninterrupted improvization

You can use a screenshot, concept art, or a gameplay video taken from any game for inspiration and context. This exercise can be harder than it seems if you have never attempted this. Remember that properly setting up everything in advance is what might allow you to really get into the zone.

Make sure you have all the instruments you would like to use easily accessible and setup the recording so it can flow completely uninterrupted. If you want to try the full Vangelis method, then all performing/mixing/producing must happen on the fly. This improvizational approach might not be right for every occasion, but it can be liberating for composers who want to get away from the constant stop/go type of writing and editing, clicking through menus, and recording fragmented ideas in a DAW environment. Afterall, there might be something magical to be discovered if we allow ourselves to get immersed deeper into our creative process.

Task 3 – Implementation (medium/challenging) – Create a 3D spatial music mix

Take all the instrumental stems of one of your pre-existing compositions and mix them as individual spatial sources within a 3D environment. You can use a free template in a game engine such as UE5 (ex: Third Person Template) and experiment with different attenuation settings and positions for each instrument. If you have never used a game engine before then the learning curve might be quite steep, but the process is relatively straightforward if you become familiar with the basics of 3D mixing.

Notes

1 Weber, "5 Things We Learned about Shadow of the Tomb Raider from the Official Lara Experts."
2 Banas and D'Oliveira, "Interview: A Chat by the Campfire with Shadow of the Tomb Raider Composer Brian D'Oliveira."
3 Banas and D'Oliveira, "Interview: A Chat by The Campfire with Shadow of the Tomb Raider Composer Brian D'Oliveira."
4 "Spatialization Overview – UE5 Documentation."
5 "Shadow of the Tomb Raider | Pinewood Studios."
6 D'Oliveira, "Brian D'Oliveira Helming Music Composition for Shadow of the Tomb Raider."
7 Remington and D'Oliveira, "Brian D'oliveira Follows Lara into the Jungle in 'Shadow of the Tomb Raider'."
8 Bridgett and D'Oliveira, *Shadow of the Tomb Raider – Sound and Music.*
9 McConnell, "'The Instrument' – Mcconnell Studios."

Bibliography

Banas, Graham, and Brian D'Oliveira. "Interview: A Chat by the Campfire with Shadow of the Tomb Raider Composer Brian D'Oliveira". *Push Square*, 2019. https://www.pushsquare.com/news/2019/01/interview_a_chat_by_the_campfire_with_shadow_of_the_tomb_raider_composer_brian_droliveira.

Bridgett, Rob, and Brian D'Oliveira. *Shadow of the Tomb Raider – Sound and Music*. Video, 2018. https://www.youtube.com/watch?v=Y6LRk7SXaE8&ab_channel=TombRaider.

D'Oliveira, Brian. "Brian D'Oliveira Helming Music Composition for Shadow of the Tomb Raider". *The Music of Tomb Raider*, 2018. http://www.musicoftombraider.com/2018/07/brian-doliveira-helming-music.html.

McConnell, Matt. "'The Instrument' – Mcconnell Studios". *Mcconnell Studios.* Accessed 2 October 2022. https://www.mattmcconnell.com/the-instrument.

Remington, Kate, and Brian D'Oliveira. "Brian D'oliveira Follows Lara into the Jungle in 'Shadow of the Tomb Raider'". *WSHU – Public Radio*, 2019. https://www.wshu.org/culture/2019-02-12/brian-doliveira-follows-lara-into-the-jungle-in-shadow-of-the-tomb-raider.

"Shadow of the Tomb Raider | Pinewood Studios". *Pinewood Studios.* Accessed 2 October 2022. https://pinewoodgroup.com/pinewood-today/credits/shadow-of-the-tomb-raider.

"Spatialization Overview – UE5 Documentation". *Unreal Engine.* Accessed 2 October 2022. https://docs.unrealengine.com/5.0/en-US/spatialization-overview-in-unreal-engine/.

Weber, Rachel. "5 Things We Learned about Shadow of the Tomb Raider from the Official Lara Experts". *Games Radar+*, 2018. https://www.gamesradar.com/5-things-we-learned-about-shadow-of-the-tomb-raider-from-the-official-lara-experts/.

Chapter 19

Control (2019)

A conversation with the composer Petri Alanko

About the composer

Petri Alanko is a BAFTA nominated Finnish composer and producer who has written music for many hit games of the Finnish game studio *Remedy Entertainment* such as *Alan Wake*, *Quantum Break*, and *Control* (composed in co-operation with Martin Stig Andersen) (Figure 19.1).

About the game

Control is a third-person action-adventure game released in 2021. It follows the story of agent Jesse Faden as she investigates the strange supernatural

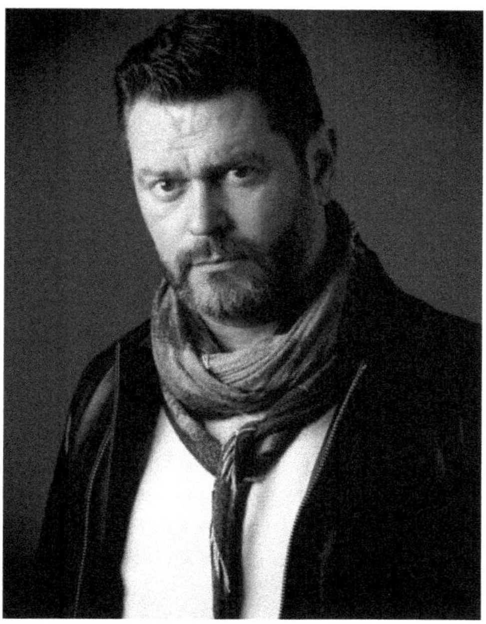

Figure 19.1 A photo of composer Petri Alanko.

DOI: 10.4324/9781003146872-20

phenomena that occur within a secret US government facility. The game was praised by fans for its innovative combat system and highly destructible environments.

Composition techniques – Sonic manipulation and found sound

MARIOS ARISTOPOULOS (MA): **Could you please describe your overall approach when composing a piece that uses found sound? Do you prefer to plan ideas in advance or is it more of an improvizatory process?**

PETRI ALANKO (PA): Oh, I tend to rely on embracing the moment. I rarely regret anything, and usually I've got a field recorder in some form with me – and to be fair, iPhone's microphone is surprisingly good for "character sound". It's my go-to nowadays. It's mostly improvization, I must say, and the sound leads, not the plan. I guess this is crucial for the result and for my own happiness – there's nothing more frustrating than trying to control a howling wind to settle down to C# when the rest of the globe wants it to be a wee bit flat F#... I know one OCD sound designer who has perfect pitch, and he tries his best to avoid field recording because of his ability. Poor lad!

MA: **I read on another interview that you relied on unusual sound sources such as an espresso machine, a microwave, and even put a piano on fire! Could you please indicate some of these on the OST so the reader can listen to how they turned out?**

PA: Oh, yeah. *Quantum Break* and *Control* suffer most from these, and something leaked over to Crossfire, too, but only a little. In QB the piano and espresso machine surface in the low range quite often – the espresso machine's "mmm-MMMMM-mmm-MMM" transformer-like low hum doubles many bass sounds. Piano is being used as a riser in, for instance, QB's "Dodging Bullets" – the background tonal field is piano being mistreated with a toothbrush and lighter fluid and a drill [video example 77, 0:40]. Piano bowed with a bottle brush is being used in the intro of "A Whisper", underneath the fragile string layer.

I recycled some of QB-era sounds with *Control*, and extended the libraries with something more aggressive [video example 78]; I went to extremes at some point and dropped a piano frame on the concrete from a forklift. I produced a nice sound, but the frame decided to land in a different angle and the contact mics got crushed in that... and, of course the floor had a crack, so it cost me quite a little to fix all that.

MA: What was your recording and editing process like? Do you ever record your sounds in sync to picture?

PA: Sometimes, but very, very rarely. I like to have a lot of material and then build something out of it. It is my preferred way to create anything – and usually, there's not much picture to record to, when I'm doing the first phase. With luck, some early placeholder cines at best, but usually still concept pictures and a screenplay only. I just dive into the ocean of imagination and try to deal with my brain. Some themes are rather happy accidents, especially when dealing with feedback – or natural overtones.

MA: Do you usually think about how an object will be processed and transformed before recording it?

PA: I'd love to say some sound designers have an ability to "see" an effect plugin/insert chain when they hear a certain right raw sound, a little like seeing a Roland SH-101's front panel and "hearing" the sound in your head. A lot like that happens with my doings: I tend to categorize raw sounds for granulators, for spectral smearing, extreme stretching etc., you get there if you have been doing editing and processing long enough, that certain kind of a "brain pathway" from a raw sound to something you need for a sound effect or a virtual instrument. I nowadays use a lot of contact microphones to catch most of the vibration – I've noticed that for some reason, contact mic sounds react best to extreme stretching. All noise gets multipled in that, so it's feasible to avoid it.

MA: Are you a fan of generative techniques such as using probabilities and randomization to vary the musical outcome?

PA: OH YES I AM! I love chaos and finding some meaning in the chaos, but usually I utilize randomization with other parameters than pitch – except for special effects, and one piano sound that tunes its strings a little off every time the key is pressed down, and of course in the higher range two strings and then three... but nothing overly random. With percussion, I tend to mimic a certain "hitting the skin in a slightly different place every time with a slightly differing force" effect every time, even with the electronic percussion, but the changes have to be really careful. Otherwise it'll sound like a badly programmed toy organ.

But, yeah, at some point I was experimenting in Kontakt scripting environment with root notes and altering the other intervals according to bass note (4th, 5th or 7th up or down), but as I'm not much good with LUA [a programming language], I gave up soon. However, I'd be willing to continue that exploration at some point, as the results were interesting.

Controlled Chaos, yes, that would be my imaginary theme park's name (Figure 19.2).

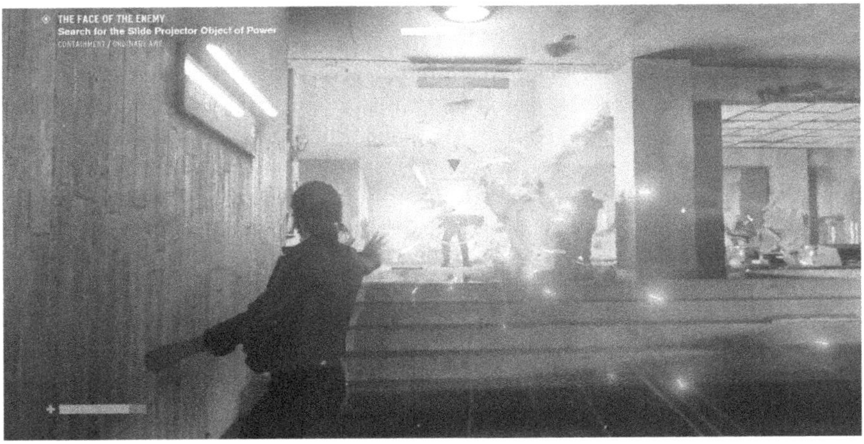

Figure 19.2 A gameplay screenshot from *Control*. The player can wreak havoc on her enemies by using telekinesis and other psychic abilities to turn the destructible environment against them.

Composition techniques – Using rule sets and interactive FX

MA: What type of interactivity is there between the game and the music?

PA: Nowadays, with WWise and FMod and Unreal5's audio engine... oh man, what an open world there is! Of course the open endedness brings its own trouble, but it also adds up to immersion, and that is something one must embrace – to catch the gamer and carry them into another world. The more we can support the gamer's actions, the better. But also: the more we can control the gamer's actions, twice better. It could work both ways, but subtlety is the key here, and it should be like a pendulum: sometimes it's the gamer that leads, sometimes it's the game. That way, the intensity is kept alive.

In *Control*, the sound designers and integrators did a huge amount of work to create a template with maybe hundreds of rule sets – "if this, then that, otherwise that and those" – and it resembles a living creature, really. Or cthulhu, actually. But the main thing is, it really reacts to the environment and the events and the action. The downside is it can sound a little Schoenbergian or serialistic, or even random, but with certain right type of a sound set it really is an effective tool.

With AI and machine learning arriving, we're soon facing something really interesting. At some point I was Slush's (the yearly Finnish startup/geek festival in the fall) Music Director, and had a chance to talk with quite a few machine learning devs about music and its role in the future, and that's something that's going to change in the upcoming years. Right now we're in the standby position, really, but the AI is there in the background, learning and running through its classes. Right now there are some AI services offering something like "build your own track with AI in a minute" and the results are garbage, but it's like with the first synths: first there were oscillators only, then arrived the filters, then came the MiniMoog topology – and after two decades, a Synclavier arrived. We're now somewhere after the MiniMoog, but Synclavier is already dawning in the horizon. I can hardly wait – I don't belong to the "fear for your profession" school, I'm more of an "I'll embrace my machine overlords" person. Maybe a composer's role will change, but to be honest, if an AI puts you out of business, there maybe was something wrong with your choice of profession in the first place. I, for instance, would gladly turn into a curator/condutor – that's a profession we're going to need when the AI strikes. They're effective, but they're initially emotionally stupid despite their endless intelligence, unless some degree of curating is conducted. And, of course, they'll learn that, too. Another layer is inventing, another judging/valuating.

MA: **Are there any FX applied in real time during gameplay or is all the audio processing rendered before implementation? I believe that some of the music appears to slow down when there are no enemies present but I might be imagining that!** [video example 79, especially from 09:30 to 10:15]

PA: There's that, you're right. It's no imagination. With *Quantum Break*, we already tested some filtering during a... teleport? Whichever is the correct term, that "thing" caused some filtering to be active during a rush or a sped-up attack. With *Control*, even more so. In *Quantum Break*, a certain Finnish individual was used for creating a granular plugin straight into the sound engine that was used for some of the inventions, but to my knowledge, that wasn't taken into action in *Control*. Whoever was the dude, I'm under the impression he was a little hard to catch. Academic doctor level people tend to have their own pacing, you see...

Usually it's Impulse Responses or convolution reverbs that are done in the playback engine, but other stuff is there, too. With music – especially during a cinematic or a pre-rendered section – there's no other processing except very occasional dynamic processing.

Production tools – Electromagnetic microphones and granular synths

MA: **Do you have any favourite gear or production tools you used in the game? I read that you used a special microphone for radiation!?**

PA: Yes! It was made by LOM and was called – I think – ElektroSluch 3+. Basically it picks up anything from electronic devices (well, electromagnetic microradiation) and amplifies that. For instance, if you put the device on your iPhone when the screen's off and call your phone from another, it'll be a majestic mayhem! Similar sound sources could be found everywhere in your home – and I actually used one for finding an electric wire in a wall before drilling, so they're very, very usable. I strongly suggest one, and please – do some time stretching or spectral smearing! [video example 80]

MA: Could you please share with us some of the sonic manipulation techniques that you used? I assume some of it must be using granular synthesis?

PA: Yes, granularity I love. I've got a few Reaktor based ensembles I created long ago that I still like to use on a daily basis, and when Waldorf Quantum arrived, I bought it immediately: it's still one of the rare hardware machines able to produce quality granular effects to be played back musically. Some ready-built granular effects or instruments are someone's fever dreams and beyond playability so badly they need to rethink their philosophy right away. What is wrong with 12 tone keyboard control? Of course, it is necessary to offer people choices and freedom, but let's just say I've tried my fair share of "playing" some "granular game changers" with a laptop touchpad and – no thanks. At some point I found a Tasty Chips GR-1, which I used for some sounds, but it's either Reaktor or Quantum for me.

Another thing totally are the Kyma sounds that can literally transform your stuff into something else. I used to use Kyma a lot, but after *Control*, it's been resting in my rack peacefully. I love it and the sounds and the algorithms, and there still is that certain type of sound that only it can bring – and then, of course, the harmonic vocoding thing, plus Tau stuff. It's, unfortunately, a tame black hole, really.

Career and creative tips from a Veteran composer

MA: The most common question from student composers is how to find work when you do not have a pre-existing relationship with a game studio. Any tips here?

PA: Be persistent. If you know you're good in something, do a good demo that leaves no discussion. There's no room for "I made this two years ago and tried to play a guitar but it was too late and then I tried playing drums but I had no money and...." demos, make it work with what you got. That turns heads and proves your point of your usability and flexibility and ingenuity. I once ran into a demo that was done with only sounds coming out of the mouth of the candidate – of course, some were processed really beyond recognition – and a certain game company employed the guy right away. Don't be in the crowd, find your expertise and stand out. Just like any career, audio and music careers depend on your self-confidence and ability to move people.

I'm willing to say one demo, be that in YouTube or Vimeo or just a lonely clip in someone's Dropbox, can change their world, but it has to be so good. The same happens with TikTok and other social media services; pop stars can be made almost overnight, and the same applies here. Just be good. It's easy to say, but it's true. Nobody lays their ears on something they've already heard a dozen times, be that Williams or Zimmer.

MA: **Any creative advice for composers that are just starting to explore found sound techniques in their work?**

PA: Just record something and try turning it into a polyphonic instrument! That's where it starts with me. I usually take the pad/longer sounds under the loop at first and the shorter ones are sure to appear!

Try making something tonal first, as noises are easy. When you deal with natural overtone series, I'm certain that the "Eureka!" moment arrives in an hour. If you're unsure, avoid noise – but in my opinion, noise can help create very, very interesting tones when stretching it to the max; it no longer behaves like noise, it becomes random tones, and with some plugins, it's easy to turn that into something more controlled. Maybe that's the key: try finding order in a chaos.

MA: **How important is it to have an agent as a game composer? Many freelance composers assume that an agent will help them find work, but my understanding is that they mainly handle contracts, press, and negotiate fees?**

PA: Well, it depends. There's the upside and the downside, and according to my experience, the companies relate better to individuals without any negotiators. Or maybe I've sold my ass too cheaply, don't know... Anyway, the agent can take some 15–20% off your certain income (usually the technical fee), but they can be of great help when it comes to agreements and rights and so on. In my case, I've got a "gentleman's agreement", where I deal with Finnish and Swedish spoken areas myself and other fields are being used through the agency – but, due to my long-lasting gig with Remedy, they've been somewhat idling lately. I'd love to see that change in the future.

If you're willing to deal with things by yourself, ask a colleague. Which was what I did, and decided to use an agent for abroad projects.

MA: Thank you Petri!

Takeaway tasks

Task 1 – Composition/production (challenging) – Compose a piece of music inspired by Control *using only found sound*

Limit yourself to using only sounds you have recorded yourself using a field recorder or even your phone. You can use any sonic manipulation techniques you want but time stretching, pitch shifting, and reverb can give you a good start for textured based sounds. I also recommend exploring granular synthesis if you have not tried it before. Logic Pro has a free granular sampler inside Alchemy and Ableton has Granulator II as a free download. For percussive based sounds you might find it easier to start with material that has clear transients.

Chapter 20

Cyberpunk 2077 (2021)

Diegetic music in Night City, riff-based composition, and the sound of sci-fi

About the game

An open world, action-adventure AAA game taking place in a dystopian Cyberpunk future. The game became infamous by the huge amount of player hype during its prolonged nine years of development that resolved into countless complaints due to the heavy technical glitches found in the initial release, which led to Sony removing it from the PlayStation store and offering full refunds. Among other things, the game has been praised for its high quality graphics (that are almost impossible to render without a powerful system) and its musical score.

Fun trivia

Johnny Silverhand, a world-famous fictional rock star that lives inside the player's mind as a cybernetic A.I. is played by Hollywood superstar Keanu Reeves!

How did the composers get the gig?

The game is scored by a collaboration of three composers: Marcin Przybyłowicz and P.T. Adamczyk were in-house composers for CD Project and had previously worked on *Witcher 3,* while Paul Leonard-Morgan had previously worked on the *Warhammer 40K* games and was brought into the project by Marcin. The material was divided according to different quest lines/areas over a three-year process.

Composition technique 1 – Diegetic music in Night City

One of the reasons that roaming in the open world of *Cyberpunk 2077* feels particularly immersive is that music often originates from within the actual game world, an approach known as diegetic or source music.

DOI: 10.4324/9781003146872-21

This technique has been used frequently in film music (ex: the cantina band scene from the original *Star Wars*) as well as in many of the games discussed in this book: Joel and Ellie in the *Last of Us* occasionally play an acoustic guitar, Link in *Zelda* plays an ocarina instrument to unlock all sorts of mysteries, street musicians in *Apotheon* play the lyre as you explore ancient Athens, and in-game characters in Assassin's Creed Syndicate sing murder ballads during important narrative moments. Such uses of diegetic music can be beneficial for multiple reasons: (1) it can enhance the sense of realism and immersion of a game world, (2) it can be used interactively with the player, (3) it can aid the storytelling, and (4) it can provide cultural information about the people who live in this virtual world.

Perhaps more than any game to date, the large world of *Cyberpunk 2077* that is set primarily in the futuristic setting of *Night City*, is filled with diegetic music: every bar, night club, car radio, and live concert venue features a plethora of original songs that is meant to be heard by the game characters as well as the player, often fusing the distinction between the two. The impressive depth of diegetic music in the game was achieved by licencing more than 157 original and diverse tracks written by several big commercial artists and bands such as *A$AP Rocky*, *Grimes*, *SOPHIE*, *Refused* and many others.[1] One of the primary ways these are experienced is through a selection of 11 radio stations that are available to play in vehicle radios while the player drives around *Night City* (Table 20.1):

The open world design of the game with its extensive playtime duration could have easily fallen victim to the common pitfalls of repetitive game soundtracks. However, the use of this car radio mechanic effectively acts like a curated exploration music playlist that can be set by players according

Table 20.1 The radio stations that the player can choose from while driving in Night City[2]

In-Game Radio Station	*Genre*
89.3 Radio Vexelstrom	90s rock
92.9 Night FM	EDM
101.9 The Dirge	Hip-hop
103.5 Radio Pebkac	Techno
88.3 Pacific Dreams	Lounge
107.3 Morro Rock Radio	Classic rock
98.7 Body Heat Radio	Pop/j-pop/k-pop
106.9 30 Principales	Latin
96.1 Ritual FM	Black/death metal
95.2 Samizdat Radio	Club music
91.9 Royal Blue Radio	Jazz
Radio off	No music

to their mood and musical taste which ensures a more variable and individualized experience. This use of diegetic music also avoids the need to add complex interactive mechanics behind its playback as it does not need to respond to changes in the action. It automatically gets switched off when the player exits the vehicle or will be replaced by underscore if a particular quest trigger requires so for dramatic reasons.

It is worth mentioning that this creative use of a diegetic car radio system that is used as exploration music is not original, it originates from the well-known *Grand Theft Auto* series that introduced the same idea. However, in *Cyberpunk* this technique is expanded upon further. The same hits from the radio stations are also played within clubs and venues, as well as by various street musicians, creating a sense of multiple cultures within the city. What is particularly interesting is that the production of the songs is adapted to emulate the acoustics of each space that it is being reproduced in. For example, when visiting the rock club *Afterlife* (video example 81) the mix is very different from the radio version and uses more reverb and different EQ to replicate the sound of large club speakers. It is also implemented directionally within the 3D environment, so the sound is spatialized accordingly as you walk around the club, a technique that further adds to the realism of the diegetic experience.

Another dimension of diegetic music occurs when you get to experience performing as the Rockstar *Johnny Silverhand* that is modelled and voiced by Keanu Reeves. After jamming on the guitar while relaxing at your sofa you go on to play in a big rock concert with *Silverhand's* band *Samurai.* What is particularly impressive here is that the game is trying to break the barrier between the player and Silverhand (the cybernetic A.I. living inside your head) through the use of an interactive performance. The chaotic live concert (video example 82) is one of the most entertaining moments in the game as it successfully encapsulates the energy of performing as a rock star on stage. This is accomplished by:

1) Using an actual rock band called *Refused* that has written original songs and lyrics from the viewpoint of this character. The band's singer even worked with a specialist vocal coach specialist to imitate the voice delivery of Keanu.
2) Creating a new version of the song to match the noisy live setting with each instrumental layer mixed from the live performer's perspective. Depending on where you look during the performance the mix is adjusted and panned dynamically (ex: if you look at the drums, they sound louder than when you look at your other band mates).
3) Giving players some interactive control over the development of the song as they can occasionally choose to either start singing, playing a raging solo or just riff along.

Music theory 101 – What is a riff?

In rock music a *riff* is a memorable short phrase or chord sequence that is constantly repeated usually in the guitars, and forms the basis of a song. It is similar to the classical concept of *ostinato* (see Chapter 9 on *Assassin's Creed*) but riff-based songs usually incorporate several riffs that alternate in a cyclical way to establish the main parts of the structure (ex: verse, bridge, chorus). A central feature of most rock songs is that the riffs themselves rarely develop but the songs still maintain a sense of progression primarily by gradual changes in the arrangement. A good example of this is the rock classic *Smoke on the Water* (video example 83) that consists of three main riffs. This is quite a different approach from using a repetitive motif as the basis of a composition within a classical context. For example, listening to Beethoven's famous fifth symphony you can quickly observe that the principal motif itself is going through significant changes as the symphony progresses in its rhythm, pitch, dynamics, phrasing, accents, and voicing.

Composition technique 2 – Riff based composition

Many of the most memorable pieces in *Cyberpunk 2077* follow a simple riff-based structure that is commonly found in classic rock and metal songs. Creating songs following this well tested riff-based formula is nothing new but what is interesting here is that the composers use simple riffs outside of a rock/pop context along with a strong use of distortion, and creative arranging techniques to create a very memorable soundtrack. These techniques can be a useful addition to your composing toolbox that are relatively easy to execute.

Riffs in Cyberpunk

Have a listen to *V's* theme (the main protagonist) in video example 84 which plays in the main menu among other moments in the game. As you can see from Figure 20.1 most of the riffs are only one or two bars long. The piece builds up over time by using the same riffs but adding more layers on top of the arrangement. It begins with bass and drums, but it gradually builds up with strings and arpeggios. However, to avoid a fully predictable linear build-up there are also sudden drops and pulls back that add an element of surprise and keep things interesting.

These riffs can also work on top of each other in a vertical arrangement as they are all based around an A minor chord with different chromatic

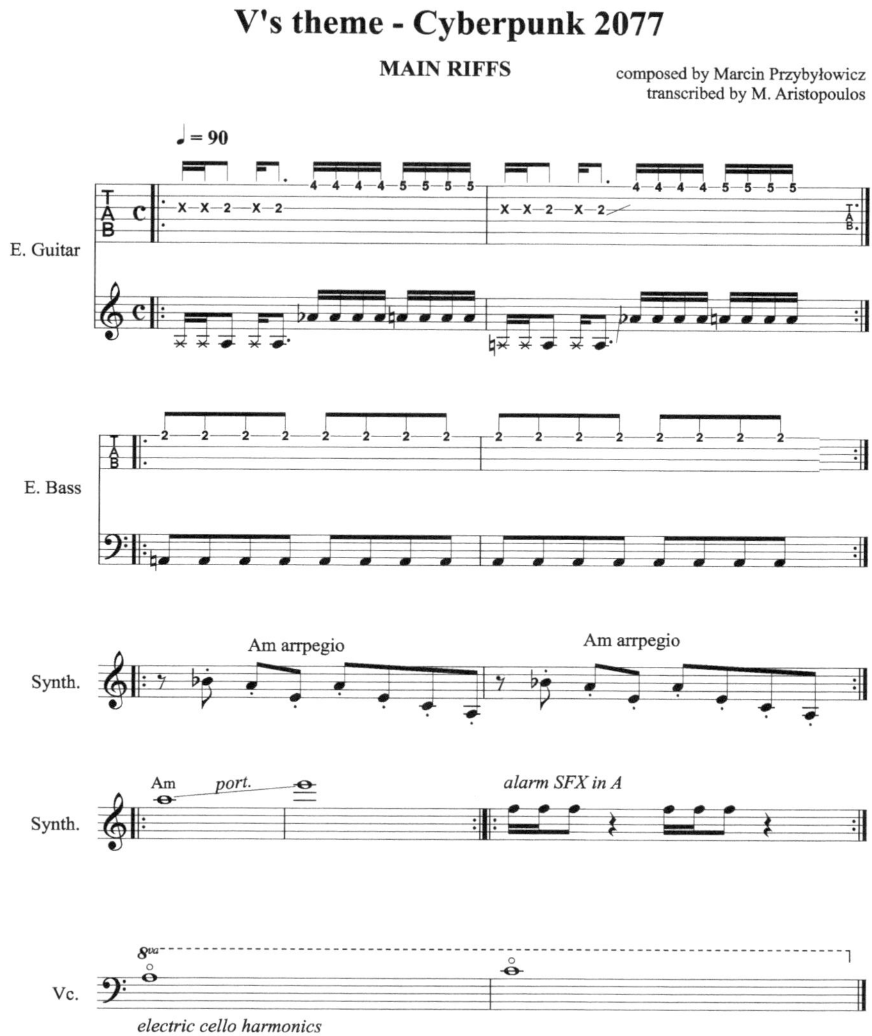

Figure 20.1 A transcription of the main riffs of *V's theme*. Notice how they are all heavily centred around the A minor chord.

passing notes. The harmonic simplicity of these riffs is especially useful in a gaming context as it is easier for the music to adapt to sudden gameplay changes and triggers without sounding jarring. This is because the number of layers can easily change without clashing as all the riffs are based on the same harmony. Moreover, any horizontal transitions with other themes are easier to plan (such as the player exiting the main menu and loading a particular save) as the harmony of this theme always remains in A minor.

The track *Musorshchiki* (video example 85) is basically made up of only two riffs. Both have a distinctive sound and are great examples of how a riff does not need to have a strong melody in order to be memorable. Riff 1 that repeats from 0:00 to 0:29 sounds almost like random radio static that glitches and stutters in a rhythmical fashion. Riff 2 that repeats from 0:30 to 0:56 plays a straight eight note pattern that uses a characteristic cyclical shift of the accents at notes 3–5–7 while simultaneously pitch bending. The song structure is very straightforward and just alternates between these two riffs but there is a number of details in the arrangement that keep it interesting:

1) The glitch rhythms of riff 1 are occasionally varied which creates a sense of unpredictability.
2) The filtering and use of production effects changes over time.
3) New details and samples are introduced at key points of the arrangement (ex: 0:29" has a short choir shout).
4) There is constant detuning and pitch shifting in the background layers that occurs at a different pace from the riff looping. This helps to avoid a structure of identical loops in each section and creates a sense of development.

The Rebel Path (video example 86) is one of the hits from the soundtrack according to the number of Spotify plays (over 5 million) and undoubtedly creates some of the most satisfying battle sequences in the game. The piece is based predominantly only on a single riff that consists of a very simple yet memorable rhythmic bass pattern of just three syncopated 16th notes that continuously loop. Observe how the use of filtering of the riff changes over time: the piece begins with most of the highs cut-off (almost resembling diegetic electronic music being heard outside a club) and develops with more aggressive harmonics and resonance being emphasized at climactic moments (ex: 0:55"). Similarly to *V's theme*, there are complimentary counter motifs that work on top of each other but stay in the same harmony (ex: lead melody at 3:16").

Composition technique 3 – Defining the sound of sci-fi

Any game that takes place in a dystopian future must inevitably answer the question of how music might sound at that time. Most often, this imaginary musical culture incorporates some version of synthesizers and electronic instruments as a means of representing technological innovation, as well as some use of traditional acoustic instruments that usually represent the human element. This electronic/acoustic duality has been long explored since the time of early sci-fi film and two iconic soundtracks that defined the genre are Bernard Herrmann's use of Theremin and orchestra in *The Day the Earth Stood Still* in 1951, and Vangelis' use of detuned analogue synth textures (ex: Yamaha CS-80) combined with ethereal vocal and saxophone melodies in *Blade Runner* (1982).

In *Cyberpunk 2077* this synth/acoustic hybrid is utilized by taking some of the main principles of rock composition (ex: riff-based arrangements, heavy distortion, pitch bending) that relate to *Jonny Silverhand* and blending them with elements of electronic, noise, orchestral, and world music to reflect the threatening yet stylish world of *Night City* and the diverse cultural backgrounds of its residents. The traditional instrumentation includes rock instruments (real drums and guitars), orchestral instruments such as the cello (which is electrified), and Japanese traditional instruments (to reflect the Japanese *Asaka* corporation heritage). The use of electronic instrumentation includes a large collection of unusual and eclectic synthesizers that have quite unique sonic signatures. These range from 1980s vintage gear such as the Soviet Formanta Polivoks Synthesizer, to new experimental modular synths such as the Folktek Mescaline (Figure 20.2). The use of retro gear is a common trend in the contemporary sci-fi genre perhaps due to those long-standing associations between synths and futurism as well as

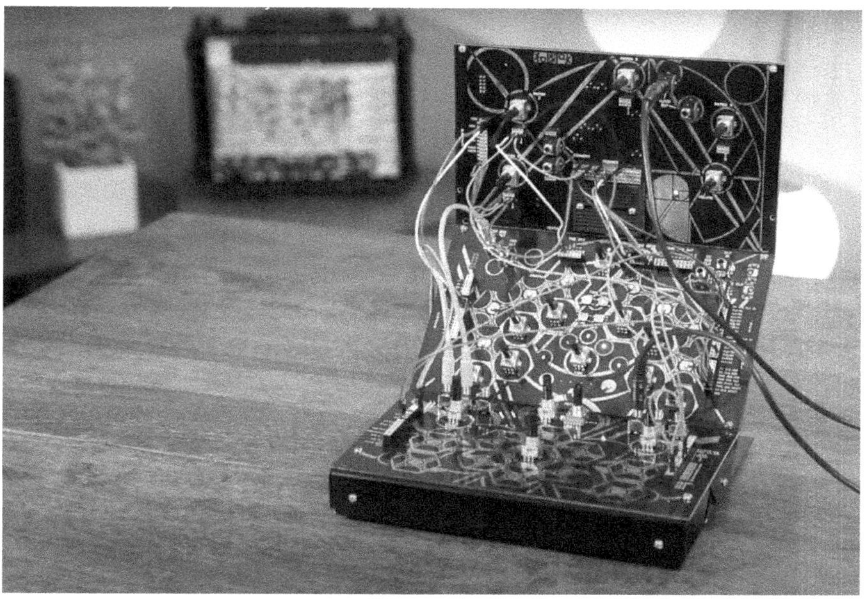

Figure 20.2 A photo of the *Folktek Mescaline*, a 10 note polyphonic synthesizer with Eurorack compatible boards that was used in the production of the *Cyberpunk 2077* soundtrack. The photograph was kindly provided by Perfect Circuit.[3]

the sense of nostalgia that is generated by using obsolete technology. This approach is just one possible direction for sci-fi instrumentation that can be modified as needed in your game music according to your own vision.

Production tools: Use of distortion and an interactive low pass filter

One of the characteristics of the production style of the game's action music (not the diegetic radio tracks) is its extensive use of heavy distortion which according to the composers is applied to every single stem of the audio! Audio distortion can be broadly defined as any alteration to an original audio signal but when it is introduced intentionally it is typically for the purpose of augmenting the frequency content of a signal in some way. The composers used long and uncommon routing paths through multiple amps, filters, and effects of analogue synths and other audio gear in order to add colour and character to the audio. For example, the output of a *moog* synth would be routed into an *Arturia Matrixbrute* synth, then into a vocoder FX, and then into a *Folktek* synth before being recorded, thus being slight distorted by each unit.[4] Moreover, they experimented with innovaitve types of distortion FX units beyond the typical guitar pedals such as the *Plasma Rack* (Figure 20.3).

> By using a High voltage step-up flyback transformer, PLASMA RACK turns your instrument's signal into a rapid series of electric discharges in a Xenon-filled tube. These powerful discharges (up to 5,500 Volts) then get picked up by a specially designed electromagnetic receiver and turned back into audio-level signal. This process results in a large amount of punishingly heavy distortion, and also saturates the sound with a wide range of harmonics and overtones.[5]

This extensive use of distortion is quickly evident by listening to any track from the game's underscore. Have a listen to the track Мусорщики again (video example 85) while looking at the very large number of harmonics and noise frequencies across the entire spectrum in Figure 20.4.

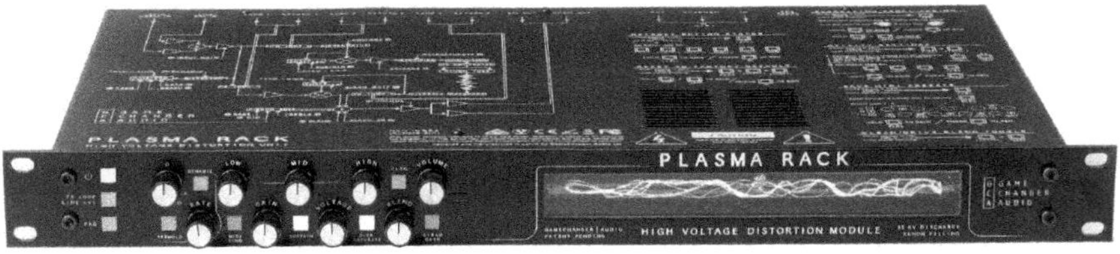

Figure 20.3 A photo of the Plasma Rack high voltage distortion effect unit used in the production of the *Cyberpunk 2077* soundtrack.[6]

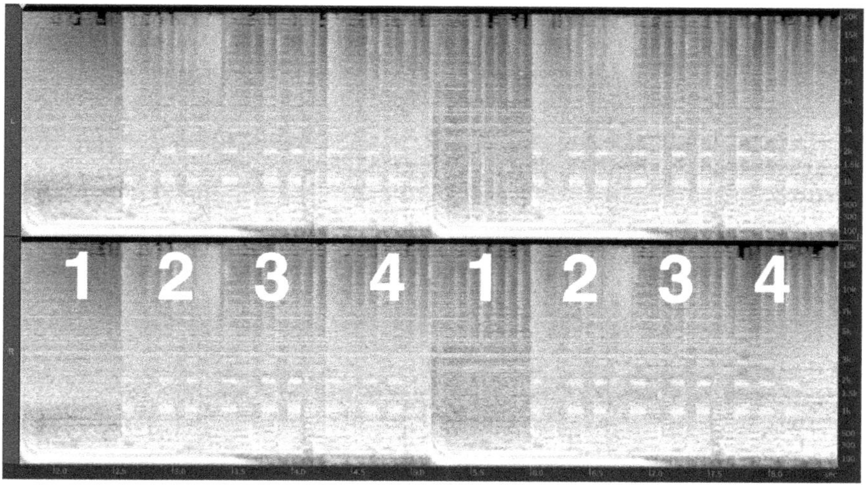

Figure 20.4 A spectrogram analysis of two bars from the track *Мусорщики*. The image provides an overview of the spectral content of the audio, with the the horizontal axis showing time and the vertical axis showing the audible frequency range (20 Hz–20 kHz).

There are three interesting points to notice here that demonstrate how the use of heavy distortion is carefully balanced and controlled:

1) There is a hole carved in the low end (below 150 Hz) so the deep bass can cut through the noise and hit strongly on beats 1 and 2.
2) Distortion is used rhythmically on the upper range with the noise stutter elements coming in primarily on beats 2, 3, and 4.
3) The snare and other percussion elements sound quite small/thin compared to the rest of the production. This was done intentionally by the composers to provide enough space for the rest of the distorted instruments to cut through the mix but also to avoid clashing with heavy battle SFX such as machine gun fire.[7]

An interactive low pass filter

In contemporary games it is quite common to apply audio effects interactively within a game engine to process SFX in real time. For example, a different reverb effect is usually applied to the same footstep sounds to match the varying acoustics of different locations. However, in some more rare cases interactive FX can also be used creatively to process elements of the music. A simple but effective way that *Cyberpunk* explores this concept is by using an interactive low pass filter in some action sequences in which the cut-off and resonance adapting to changes in the gameplay tension. One of the most impressive uses of this technique that works very effectively in increasing the feeling of synchronization and immersion can be observed in video example 87 (24:15–25:45). This video demonstrates a gameplay capture of an early mission in which the player must locate and rescue a particular character within a heavily guarded apartment. As the player starts sneaking across the room, observe how the high end of the music is filtered out when the threat of getting capture is low, but as the risk of getting caught increases by moving close to an enemy, the filter dynamically opens to a higher range, allowing the music to become more piercing and aggressive just at the perfect moments in the action.

Takeaway tasks

All three tasks below can be done individually or combined.

Task 1 – Composition (medium) – Write a diegetic theme that originates from Night City

Pick a location, character, or context within the game that you find interesting. You can look at a night drive gameplay capture of the city for inspiration (video 88). Also, think about how it could be implemented into the game and adjust your production approach accordingly. Is it playing through speakers in a club? Is it a musical performance by an NPC? Is there an instrument that the player interacts with?

Task 2 – Composition (easy/medium) – Write a riff-based theme

You might find it fun and easy to compose a piece of music based on short riffs rather than try to develop longer sections of music. Remember that riffs do not have to be used only in a rock guitar context but can be applied to any genre, and that the sense of development might have to come through creative arranging and production techniques.

Task 3 – Production (medium/challenging) – Produce a track in which you explore different uses of distortion on every instrumental layer

You do not need to own an exotic collection of analogue gear, you can experiment with distortion FX from a variety of sources (ex: pedalboard, audio inserts, synth effects, external racks, unusual mixing paths in your DAW, Re-wire, etc.). Remember that FX of the same type (ex: overdrive distortion) that come from different sources are still likely to have their own individual sound characteristics as they alter the audio signal in a different way (ex: adding a different ratio of overtones). You might be surprised to find interesting FX within a cheap synth plug-in that you otherwise do not like. Also, make sure to try to shape the distortion over time and balance it with the rest of your stems by using filters and/or FX automation, otherwise this task might quickly add up to a noisy mess!

Notes

1 Anderson, "Here Is Every Song in the Soundtrack for Cyberpunk 2077."
2 "Cyberpunk 2077: In-Game Music Credits, All the Songs Listed!."
3 "Music Technology & Synthesizers – Perfect Circuit."
4 Ruppert et al., "Sitting Down with the Composers Behind Cyberpunk 2077's Soundtrack."
5 "PLASMA Rack."
6 "PLASMA Rack."
7 Williams et al., "An Interview with the Composers for CYBERPUNK 2077."

Bibliography

Anderson, Maia. "Here Is Every Song in the Soundtrack for Cyberpunk 2077". *Mxdwn Music*, 2020. https://music.mxdwn.com/2020/12/11/news/here-is-every-song-in-the-soundtrack-for-cyberpunk-2077-including-asap-rocky-health-metz-and-more/.

"Cyberpunk 2077: In-Game Music Credits, All the Songs Listed!". *Soundtracks, Scores and More*, 2020. https://soundtracksscoresandmore.com/2020/12/09/cyberpunk-2077-in-game-music-credits-all-the-songs-listed/.

"Music Technology & Synthesizers – Perfect Circuit". *Perfect Circuit*. Accessed 15 October 2022. https://www.perfectcircuit.com/.

"PLASMA Rack". *Gamechanger Audio*. Accessed 2 October 2022. https://gamechangeraudio.com/plasma-rack/.

Ruppert, Liana, Marcin Przybyłowicz, P.T. Adamczyk, and Paul Leonard-Morgan. "Sitting Down with the Composers Behind Cyberpunk 2077's Soundtrack". *Game Informer*, 2020. https://www.gameinformer.com/2020/11/19/sitting-down-with-the-composers-behind-cyberpunk-2077s-soundtrack.

Williams, Tommy, Marcin Przybylowicz, P.T. Adamczyk, and Paul Leonard-Morgan. "An Interview with the Composers for CYBERPUNK 2077". *Geek Tyrant*, 2020. https://geektyrant.com/news/an-interview-with-the-composers-for-cyberpunk-2077.

Index

Note: Page numbers followed by "n" denote endnotes.

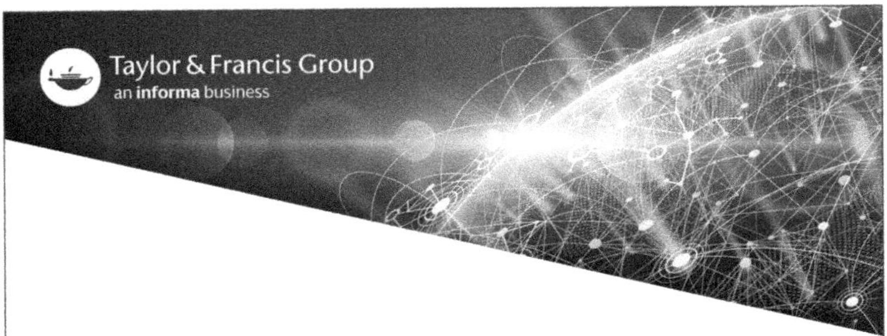

Taylor & Francis eBooks

www.taylorfrancis.com

A single destination for eBooks from Taylor & Francis with increased functionality and an improved user experience to meet the needs of our customers.

90,000+ eBooks of award-winning academic content in Humanities, Social Science, Science, Technology, Engineering, and Medical written by a global network of editors and authors.

TAYLOR & FRANCIS EBOOKS OFFERS:

- A streamlined experience for our library customers
- A single point of discovery for all of our eBook content
- Improved search and discovery of content at both book and chapter level

REQUEST A FREE TRIAL

support@taylorfrancis.com

Routledge Taylor & Francis Group | CRC Press Taylor & Francis Group

For Product Safety Concerns and Information please contact our EU representative GPSR@taylorandfrancis.com
Taylor & Francis Verlag GmbH, Kaufingerstraße 24, 80331 München, Germany

www.ingramcontent.com/pod-product-compliance
Ingram Content Group UK Ltd.
Pitfield, Milton Keynes, MK11 3LW, UK
UKHW020005200926
475702UK00015B/2494
* 9 7 8 0 3 6 7 7 0 5 4 9 7 *